ISNM 102:
International Series of Numerical Mathematics
Internationale Schriftenreihe zur Numerischen Mathematik
Série Internationale d'Analyse Numérique
Vol. 102

Edited by
K.-H. Hoffmann, Augsburg; H. D. Mittelmann, Tempe;
J. Todd, Pasadena

Birkhäuser Verlag
Basel · Boston · Berlin

Random Partial Differential Equations

Proceedings of the Conference held at the Mathematical Research Institute at Oberwolfach, Black Forest, November 19–25, 1989

Edited by

U. Hornung
P. Kotelenez
G. Papanicolaou

1991

Birkhäuser Verlag
Basel · Boston · Berlin

Editors

Prof. Dr. U. Hornung
SCHI
P.O. Box 1222
D–W–8014 Neubiberg
Germany

Prof. P. Kotelenez
Department of Mathematics
and Statistics
Case Western Reserve University
Cleveland, OH 44106
USA

Prof. G. Papanicolaou
Courant Institute
251 Mercer Street
New York, NY 10012
USA

Library of Congress Cataloging-in-Publication Data

Random partial differential equations : proceedings of the conference
held at the Mathematical Research Institute at Oberwolfach, Black
Forest, November 19–25, 1989 / edited by U. Hornung, P. Kotelenez, G. Papanicolaou.
(International series of numerical mathematics =
Internationale Schriftenreihe zur numerischen mathematik = Série
internationale d'analyse numérique : vol. 102)
Papers from the Conference on "Random Partial Differential Equations."
ISBN 3-7643-2688-3 – ISBN 0-8176-2688-3
1. Stochastic differential equations–Congresses. 2. Differential
equations, Partial–Congresses. I. Hornung, Ulrich. 1941– .
II. Kotelenez, P. (Peter), 1943– . III. Papanicolaou, George, 1943–
IV. Conference on "Random Partial Differential Equations" (1989 :
Mathematic Research Institute at Oberwolfach) V. Series:
International series of numerical mathematics ; v. 102.
QA274.R355 1991
519.2 – dc20

Deutsche Bibliothek Cataloging-in-Publication Data

Random partial differential equations: proceedings of the
conference held at the Mathematical Research Institute at
Oberwolfach, Black Forest, November 19–25, 1989 / ed by
U. Hornung . . . – Basel ; Boston ; Berlin : Birkhäuser, 1991
(International series of numerical mathematics ; Vol. 102)
ISBN 3-7643-2688-3 (Basel . . .)
ISBN 0-8176-2688-3 (Boston)
NE: Hornung, Ulrich [Hrsg.]; Mathematisches Forschungsinstitut
<Oberwolfach>; GT

Printed from the authors' camera-ready manuscripts on acid-free paper in Germany
ISBN 3-7643-2688-3
ISBN 0-8176-2688-3

Preface

The present volume contains manuscripts of lectures which were given at the conference on "Random Partial Differential Equations" at the Mathematical Research Institute at Oberwolfach. The conference took place during the week of November 19, 1989, and was organized by the editors. Researchers in probability and differential equations were invited and 37 scientists from 12 countries participated. Topics of the papers presented at the conference were parabolic Itô equations, Malliavin calculus, Dirichlet forms, Markov fields, Skorohod and Hellinger integrals, random Schrödinger operators, Lyapunov exponents, and optimal control of stochastic systems. Furthermore, limit theorems of particle systems, acoustic waves in random media, and random oscillations were discussed.

The thirteen papers in this volume are representative of the topics presented in the lectures.

The papers of Chow-Jiang, Jetschke and Kotelenetz are on the analysis of stochastic partial differential equations, primarily with tools from functional analysis. The papers of Grigelionis and Millet-Nualart-Sanz and Manthey rely more on probabilistic ideas and methods. The paper of Golden is on the use of random partial differential equations in composites and on homogenization methods. The paper of Benassi-Fouque is on the hydrodynamic behavior of interacting particle systems where both probabilistic and analytical techniques are used. The paper of Wedig is more applied in character and relates directly to problems in mechanics. The papers of Kirsch-Pastur, Dell'Antonio, and Alberverio-Blanchard-Ma are motivated from problems in mathematical physics and use a variety of analytical and probabilistic tools; they deal with Schrödinger's equation and Dirichlet forms.

The editors wish to thank Prof. M. Barner and his co-workers of the Mathematical Research Institute for their help in creating the conference. The editors also wish to thank Birkhäuser Verlag for their production of this book which will enable many more scientists to enjoy the stimulation of the conference.

Ulrich Hornung (Neubiberg)
Peter Kotelenetz (Cleveland)
George Papanicolaou (New York)

Contents

List of Contributors

Sergio A. ALBERVERIO, Fakultät für Mathematik, Ruhr-Universität, W-4630 Bochum 1, Germany

Pao-Liu CHOW, Department of Mathematics, Wayne State University, Detroit, Michigan 48202 U.S.A.

Gianfausto F. DELL'ANTONIO, Dipartimento di Matematica, Univ. di Roma I, P. Aldo Moro 2, I-00185 Roma, Italy

Jean-Pierre FOUQUE, CNRS, Centre de Mathématique Appliquées, Ecole Polytechnique, F-91128 Palaiseau Cedex, France

Kenneth GOLDEN, Deptartment of Mathematics, Princeton University, Princeton, NJ 08544 U.S.A.

Luis G. GOROSTIZA, Centro de Investigación y de Estudios Avanzados, México 07000 D.F., México

B. GRIGELIONIS, Institute of Mathematics and Cybernetics, 232600 Vilnius, Akademijos 4, Lithuania

G. JETSCHKE, Institut für Ökologie der Universität, Neugasse 23, O-6900 Jena, Germany

Werner KIRSCH, Fakultät für Mathematik, Ruhr-Universität, Postfach 102148, W-4630 Bochum 1, Germany

Peter KOTELENEZ, Deptartment of Mathematics, Case Western Reserve University, Cleveland, OH 44106 U.S.A.

Ralf MANTHEY, Friedrich-Schiller-Universität, Sektion Mathematik, UHH 17. OG, O-6900 Jena, Germany

David NUALART, Universitat de Barcelona, Facultat de Matemàtiques, 08071-Barcelona, Spain

W. WEDIG, Institut für Technische Mechanik der Universität, Kaiserstr. 12, W-7500 Karlsruhe 1, Germany

Feynman-Kac semigroups in terms of signed smooth measures

by

Sergio Albeverio[*,**,***], Philippe Blanchard[**,#], ZhiMing Ma[**,##]

ABSTRACT

We study positive continuous additive functionals of diffusion processes associated with Dirichlet forms. We also give necessary and sufficient conditions for a Feynman-Kac semigroup associated with a signed Borel measure to be a strongly continuous semigroup in the relevant L^2-space. We characterize their generators as second order partial differential operators with zero order term (potential) given by a signed measure (which can be so singular as to be nowhere Radon). We also discuss preservation of p-boundedness and L^p-smoothing properties of semigroups under perturbations. We also study integral kernels of Feynman-Kac semigroups and provide upper bounds for them.

* Fakultät für Mathematik, Ruhr-Universität, D-4630 BOCHUM 1 (FRG)

** BiBoS Research Center, D-4800 Bielefeld (FRG)

Fakultät für Physik, Universität Bielefeld

On leave of absence from Academia Sinica, Beijing

*** SFB 237 - Essen - Bochum - Düsseldorf; CERFIM, Locarno

1. Introduction

Classical and axiomatic potential theories have been formulated originally in spaces of continuous functions and their duals. The relations of those "C–theories" and classical diffusion processes are well known and one of the big discoveries of our century. Feller's theory of generalized diffusion operators in one dimension can be looked upon as an extension of those relations to a larger class of generators (allowing in particular strong singularities). The theory of Dirichlet forms, as initiated by Beurling–Deny and pursued particulary by Fukushima and Silverstein [F1], [S1] provides a natural extension of Feller's theory to the multidimensional case, at least when the generators are symmetric (which corresponds to time reversible processes).

It was realized in [AHK] and [AHS] that this theory is ideally suited to handle the Schrödinger operators arising in quantum theory (both in the case of finitely many and infinitely many degrees of freedom) in particular those which have strong singularities, see e.g. [ABRö], [AFHKL], [AGKS] and references therein. This makes itself manifest in the definition of the operators, the associated semigroups, unitary time evolution groups and processes, as well as in developing methods of approximation, see e.g. [ABRö], [AKS] and references therein. For estimating spectral quantities like eigenvalues and eigenfunctions the writing of Schrödinger operators as a "free, kinetic part" and an "interaction, potential part" is useful and a probabilistic goal for it is the derivation of a Feynman–Kac formula, see e.g. [Sim1] and references therein. The study of the Feynman–Kac formula and its connection with classical analysis was initiated in the late forties. In the last decade there has been a revival of the discussion of this formula, using either probabilistic or analytic tools, see e.g. [ASi], [Br], [AFHKL], [BrT], [Che], [DC], [Pa], [BHH], [Ma], [MaS], [Ch], [J], [Sim2], [Zh], [St]. More generally, Feynman–Kac functionals can be defined as continuous additive functionals (PCAF) of Brownian motion, and in fact of any process associated with a (regular) Dirichlet form.For discussions of such functionals see [BM1-3], [AM2-5]. It was discovered recently [AM2] that there are Feynman–Kac functionals given by smooth measures in the sense of [F1] which are nowhere Radon.On the other hand it was also shown that all Feynman–Kac functionals, also those associated with nowhere Radon smooth measures, can be approximated in a suitable sense, by Feynman–Kac functionals in a certain class, called Kato class [AM4].In the case of functions instead of measures, Kato class is the one discussed in [ASi]. For applications it is important to have perturbations formulas relating a given class of Feynman–Kac formulas to other "simpler" ones. More generally, it is interesting to have a perturbation theory of semigroups associated with Dirichlet forms by potentials which are signed smooth measures. Such a perturbation theory was presented first in [AM4] giving among other things improvements of the KLMN theory (cf. [RS1,2]). Many other properties of (generalized) Schrödinger operators and associated semigroups have been discussed in the literature, and can be extended to our setting in terms of perturbation theory of Dirichlet forms. In particular the L^p–smoothing property of Schrödinger operators is discussed in [Sim] and [V] and we discuss in our paper how it can be extended to the case of perturbations given by signed smooth measures. Integral kernels of semigroups given by Feynman–Kac functionals associated with signed measures in Kato class have been discussed in e.g. [C] and [Sim]. In this paper we extend the discussion to our more general setting and provide also estimates of the integral kernel in terms of the heat kernel i.e. the integral kernel associated with the Wiener process.

This paper is organized as follows. In Section 2 we describe some assumptions on the underlying diffusion process associated with regular Dirichlet forms, and give some examples which frequently appear in the mathematical and physical literature, and we remark that they verify our assumptions. In Section 3 we recall some basic facts about positive continuous additive functionals and smooth measures in terms of the $\mathcal{E}$–nests introduced in our recent work [AM1]. This section is also devoted to discussions on the smooth measures in Kato class which were introduced in [BM1] for Brownian motion and in [AM3] for general cases. In Section 4 we give necessary and sufficient conditions for a Feynman–Kac semigroup to be a strongly continuous semigroup in the relevant L^2 space. In the classical case the generator of the Feynman–Kac semigroups are exactly Schrödinger operators (in distributional sense) with potentials given by signed smooth measures, which may be singular on each neighborhood of every point x (this is due to the discovery of nowhere Radon smooth measures, see [AM3]). As a preparation for discussing L^p–smoothing, in Section 5 we describe compatibility between two smooth measures and give examples to illustrate it. Under the compatibility assumption, we prove in Section 6 that the Feynman–Kac semigroup inherits the L^p–boundedness and the L^p- smoothing properties of the original semigroup. This result extends previous results (e.g. in [Sim]) to more general semigroups as well as to more general potentials. In Section 7 we study the integral kernels of Feynman–Kac semigroups, for which we give upper bounds in Section 8.

2. Assumptions and examples

$(\mathcal{E},\mathcal{F})$ is a regular Dirichlet form on $L^2(X,m)$, where X is a locally compact separable metric space and m is a positive Radon measure on X with supp $[m] = X$. $(X_t) := (\Omega, X_t, \zeta, P_x)$ is the unique (up to equivalence) Hunt process which generates $(\mathcal{E},\mathcal{F})$, i.e. $\mathcal{F} = \mathcal{D}(\sqrt{-H})$ and $\mathcal{E}(f,g) = (\sqrt{-H}f, \sqrt{-H}g), \forall f,g \in \mathcal{D}(\sqrt{-H})$. Here and henceforth, H denotes the L^2–generator of (X_t) and $\mathcal{D}(A)$ always denotes the domain of an arbitrary operator A. Let $(P_t)_{t>0}$ and $(R_\alpha)_{\alpha>0}$ be respectively the transition function and the resolvent of (X_t) . $\mathcal{B}_b(X)$ and $C_b(X)$ are the sets of bounded Borel functions and bounded continuous functions on X, respectively.

We shall sometimes impose the following additional assumptions.

2.1 Assumption $\quad P_t f \in C_b(X), \forall f \in \mathcal{B}_b(X), t > 0.$
Obviously assumption 2.1 implies that

$$P_t(x,\cdot) << m \quad , \quad \forall t > 0 \text{ and } x \in X \; . \tag{2.1}$$

$<<$ meaning absolutely continuity. The transition density function of $(P_t)_{t>0}$ will be denoted by $p(t,x,y)$.

2.2 Assumption P_t is a bounded operator from $L^1(X,m)$ to $L^\infty(X,m)$ for each $t>0$.
Assumption 2.2 in fact implies that P_t is a bounded operator from $L^p(X,m)$ to $L^{p'}(X,m)$ for any $1 \le p \le p' \le \infty$. See the proof of Th. 6.1(iii).
Let us denote by $\|\cdot\|_{p,q}$ the operator norm from $L^p(X,m)$ to $L^q(X,m)$. A stronger assumption than Assumption 2.2 is the following

2.3 Assumption There exist positive constants c and d such that

$$\|P_t\|_{1,\infty} \le ct^{-\frac{d}{2}}, \forall 0 < t \le 1 \tag{2.2}$$

In applications the constant d usually coincides with the dimension of the underlying space X. Notice that (2.2) implies

$$\|P_{t+1}\|_{1,\infty} \le \|P_t\|_{1,1}\|P_1\|_{1,\infty} \le c, \forall t \ge 0$$

since $\|P_t\|_{1,1} \le 1$.

2.4 Assumption
(i) $(P_t)_{t>0}$ admits an integral kernel $p(t,x,y)$ which is jointly continuous on $(0,\infty)\times X \times X$.
(ii) $P_t1(x)$ is jointly continuous on $(0,\infty)\times X$. Here 1 stands for the constant function taking value 1.

In the case where X is precisely the Euclidean space $\mathbb{R}^d$, one often meets the following stronger assumption which implies all the above assumptions 2.1 – 2.4.

2.5 Assumption $(P_t)_{t>0}$ admits a continuous integral kernel $p(t,x,y)$ such that for some constants $a,b>0$,

$$p(t,x,y) \le a(2\pi t)^{-\frac{d}{2}} \exp\left(\frac{-b|x-y|^2}{2t}\right). \tag{2.3}$$

Let us now describe some examples verifying the above assumptions. Since our main interest lies in those cases where X is locally compact let us consider the case where X is a domain D of $\mathbb{R}^d$ and m is a everywhere dense Radon measure on D. A general expression for a regular Dirichlet form of diffusion type is given by the Beurling–Deny formula [F1]:

$$\mathcal{E}(u,v) = \sum_{i,j=1}^{d} \int_X \frac{\partial u}{\partial x_i}\frac{\partial v}{\partial x_j}\nu_{ij}(dx), \quad \forall u,v \in \mathcal{F} \tag{2.4}$$

where ν_{ij}, $1 \le i,j \le d$ are Radon measures on D which are nonnegative definite and symmetric in the sense that for any $\xi \in \mathbb{R}^d$ and any compact set $K \subset D$,

$$\sum_{i,j=1}^{d} \xi_i\xi_j\nu_{ij}(K) \ge 0, \quad \nu_{ij}(K) = \nu_{ji}(K), \quad \forall 1 \le i,j \le d; \tag{2.5}$$

and where the domain $\mathcal{F}$ is the closure of $C_0^\infty(X)$ with respect to the norm $\mathcal{E}_1(u,u) := \mathcal{E}(u,u)+(u,u)$. (Of course, in order that the closure $\mathcal{F}$ exists and is contained in $L^2(X,m)$, i.e., that the form $\mathcal{E}$ initially defined on $C_0^\infty(X)$ is closable supplementary conditions must be imposed on ν_{ij}. A necessary and sufficient condition, which is called Hamza condition, for $\mathcal{E}$ to be closable has been given in [ARö2], where the closability for pre–Dirichlet form in infinite dimensional spaces has also been discussed. In this paper we are only interested in the closable case.)

The most important and simplest example is the classical Dirichlet form on $I\!R^d$, which is at the very origin of the theory of Dirichlet forms. The example is described as follows.

2.6 Example Let $X = I\!R^d$ and let m be the Lebesgue measure. The Dirichlet form $(\mathcal{E},\mathcal{F})$ on $L^2(X,m)$ is defined by

$$\begin{cases} \mathcal{E}(u,v) = \frac{1}{2}\int \nabla u \cdot \nabla v m(dx), \quad \forall u,v \in \mathcal{F}, \\ \mathcal{F} = H^1(I\!R^d) \end{cases} \tag{2.6}$$

Here $H^1(I\!R^d)$ is the Sobolev space, i.e.,

$$H^1(I\!R^d) = \{u \in L^2(X,m) : \frac{\partial u}{\partial x_i} \in L^2(X,m),\ 1 \le i \le d\}$$

where $\frac{\partial u}{\partial x_i}$ is the derivative in distributional sense. The unique diffusion process associated with $(\mathcal{E},\mathcal{F})$ is the Brownian motion (X_t) on $I\!R^d$. The L^2–generator H of (X_t) is the self adjoint extension of $\frac{1}{2}\Delta$, Δ being the Laplacian. It is well known that in this case $(P_t)_{t>0}$ admits a jointly continuous integral kernel which is given by

$$\bar{p}(t,x,y) = (2\pi t)^{-\frac{d}{2}} e^{-\frac{|x-y|^2}{2t}} \tag{2.7}$$

Consequently in this case $(P_t)_{t>0}$ enjoys all the assumptions 2.1 – 2.5.

In the following we shall always refer to the Dirichlet form described in the above example as <u>the classical Dirichlet form on $I\!R^d$</u>. A more general case is described in the following example.

2.7 Example $X = I\!R^d$ and m is the Lebesgue measure.

$$\begin{cases} \mathcal{E}(u,v) = \frac{1}{2}\sum_{i,j=1}^{d} a_{ij}(x)\frac{\partial u}{\partial x_i}\frac{\partial v}{\partial x_j} m(dx), \quad u,v \in \mathcal{F} \\ \mathcal{F} = \overline{C_0^\infty(I\!R^d)}^{\mathcal{E}_1} \end{cases}$$

where $\{a_{ij}\}$ is a symmetric matrix with bounded, Lipschitz–continuous elements and which is moreover strictly positive definite, i.e., there exists $c > 0$ such that

$$\sum_{i,j=1}^{d} a_{ij}(x)\xi_i\xi_j \geq c\|\xi\|^2, \quad x, \xi \in \mathbb{R}^d. \tag{2.8}$$

$\overline{C_0^\infty(\mathbb{R}^d)}^{\mathcal{E}_1}$ denotes closure of $C_0^\infty(\mathbb{R}^d)$ in the $\mathcal{E}_1$–norm, i.e. the norm given by the natural scalar product $\mathcal{E}_1(u,v) = \mathcal{E}(u,v) + (u,v)$ (with (u,v) the $L^2(m)$ scalar product). In this case the Markov semigroup $(P_t)_{t>0}$ admits a jointly continuous integral kernel satisfying (2.3), and hence (P_t) verifies all the assumptions 2.1 – 2.5. The generator of (X_t) is a self–adjoint extension of the following differential operator:

$$H = \frac{1}{2}\sum_{i,j=1}^{d} \frac{\partial}{\partial x_i} a_{ij}(x) \frac{\partial}{\partial x_i}. \tag{2.9}$$

See [Fr] pp. 114 for a discussion of this case. We emphasize that due to the strong positive definite property (2.8), $\mathcal{F}$ is again the Sobolev space $H^1(\mathbb{R}^d)$.

2.8 Example $X = \mathbb{R}^d$ and $m(dx) = \rho(x)dx$ where $\rho \in L^1_{loc}(\mathbb{R}^d)$ is a positive function satisfying Hamza condition (e.g. ρ is lower semicontinuous, see [ARö2])

$$\begin{cases} \mathcal{E}(u,v) = \frac{1}{2}\int \nabla u \cdot \nabla v m(dx), \\ \mathcal{F} = \overline{C_0^1(\mathbb{R}^d)}^{\mathcal{E}_1} \end{cases}$$

In this case the associated diffusion process (X_t) is called a <u>distorted Brownian motion</u> (see [F3], [AHS]). When ρ is nice enough, the $L^2(\mathbb{R}^d, m)$–generator of (X_t) is the Friedrichs extension of the following partial differential operator:

$$H = \frac{1}{2}\Delta + (\frac{1}{2}\nabla \log \rho) \cdot \nabla \tag{2.10}$$

In many practical situations the transition semigroups of distorted Brownian motions enjoy nice properties. We give a concrete example below to illustrate this point.

2.9 Example Let α be a real number, and $\rho(x) = e^{\alpha|x|^2}$ in the above example. The differential operator (2.10) will become

$$H_\alpha := \frac{1}{2}\Delta + \alpha x \cdot \nabla \tag{2.11}$$

which, for $\alpha < 0$, exactly corresponds to the <u>harmonic oscillator</u> in the physical literature (see also [AHS]). The associated diffusion process is called a <u>Ornstein–Uhlenbeck velocity</u>

process, which is, from a physics point of view, a path integral for the harmonic oscillator, and from a probabilistic point of view, a unique (up to changes of scale) invariant Gaussian Markov process. The corresponding semigroup $(P_t)_{t>0}$ admits a jointly continuous integral kernel which can be written as

$$p_\alpha(t,x,y) = \bar{p}\left(\frac{e^{2\alpha t}-1}{2\alpha}, e^{\alpha t}x, y\right) e^{-\alpha|x|^2}, \tag{2.12}$$

where $\bar{p}(t,x,y)$ is specified by (2.7). (Notice that the above integral kernel is given with respect to the measure $m(dx) := \rho(x)dx = e^{\alpha|x|^2}dx$. The corresponding integral kernel w.r.t. the Lebesgue measure is obtained by dropping the factor $e^{-\alpha|x|^2}$).

It is evident from (2.12) that, if $\alpha \leq 0$, then $(P_t)_{t>0}$ verifies the assumptions 2.1 – 2.5. A detailed discussion of the above example is done in [Sim2] and [BCZ].

2.10 Example Let $X = (0,\infty)$ and $m(dx) = x^{n-1}dx$.

$$\begin{cases} \mathcal{E}(u,v) = \dfrac{1}{2}\displaystyle\int_{(0,\infty)} \nabla u \cdot \nabla v m(dx), \\ \mathcal{F} = \overline{C_0^\infty(X)}^{\mathcal{E}_1} \end{cases}$$

In this case the associated diffusion process (X_t) is called the n–dimensional Bessel process, because its generator is nothing but Friedrichs' extension of the following Bessel operator

$$H = \frac{\Delta}{2} + \frac{n-1}{x}\nabla$$

From the probabilistic point of view, if (W_t) is an n–dimensional Brownian motion, then the process $X_t := |W_t|$ is an n–dimensional Bessel process. The corresponding semigroup $(P_t)_{t>0}$ admits a jointly continuous integral kernel given explicitly by

$$p(t,x,y) = 2C_n(2\pi t)^{-\frac{n}{2}} e^{-\frac{|x-y|^2}{2t}} \tag{2.13}$$

where C_n is a normalization constant chosen such that $\int p(t,x,y)m(dx) = 1$. From (2.13) it is easy to see that (P_t) satisfies Assumptions 2.1 – 2.5. For a detailed discussion of Bessel processes see [Mc].

Before concluding this section we impose one more additional assumption which will be used in obtaining the local continuity of the integral kernels in Section 7.

For an arbitrary set $G \subset X$, we denote by τ_G the first exit time from G i.e.

$$\tau_G = \inf\{t > 0 : \; X_t \in X - G\} \tag{2.14}$$

2.11 Assumption let G be an arbitrary open set of X and $f \in \mathcal{B}_b(X)$. Then $E_x f(X_{\tau_G})$ is continuous in G.

One can show that if (P_t) admits a continuous integral kernel $p(t,x,y)$, then a sufficient condition for Assumption 2.11 can be formulated in the following way:

$$\lim_{t \downarrow 0} \sup_{|x-y| \geq \delta} p(t,x,y) = 0, \quad \forall \delta > 0 \tag{2.15}$$

From (2.15) it follows that Assumption (2.11) is satisfied in the Examples 2.6, 2.7, 2.9, 2.10.

3. PCAF's and smooth measures

Let Cap be the 1–capacity associated with $(\mathcal{E}, \mathcal{F})$. Following [F1], a positive continuous additive functional (shortly denoted in the following by PCAF) of (X_t) is by definition an extended real valued process $(A_t)_{t \geq 0}$, which is adapted to the completed filtration $(\mathcal{F}_t)_{t \geq 0}$ generated by $(X_t)_{t \geq 0}$, such that there exists a defining set $\Lambda \in \mathcal{F}_\infty$ and an exceptional set $N \subset X$ satisfying the following properties:

(i) $Cap(N) = 0$ and $P_x(\Lambda) = 1$ for all $x \in X - N$;

(ii) $\Theta_t \Lambda \subset \Lambda$ for all $t > 0$ (Θ_t denotes the shift operator on Ω);

(iii) for each $\omega \in \Lambda$, $A.(\omega)$ is a nonnegative continuous function on $[0, \infty]$, finite on $[0, \zeta(\omega))$, vanishing at 0 and additive in the sense that $A_{t+s}(\omega) = A_t(\omega) + A_s(\Theta_t \omega)$ for all $s, t \geq 0$.

We shall denote by A_c^+ the set of all PCAF's of (X_t).

Two PCAF's (A_t) and (B_t) are said to be equivalent if $P_x(A_t = B_t, \forall t \geq 0) = 1$ for q.e. $x \in X$.
Given $(A_t) \in A_c^+$, there exists a unique Borel measure μ on X such that

$$\lim_{t \downarrow 0} \frac{1}{t} E_{h \cdot m} \left[\int_0^t f(X_s) dA_s \right] = \int_X h(x)(f \cdot \mu)(dx) \tag{3.1}$$

for all α–excessive functions $h, \alpha \geq 0$, and $f \in \mathcal{B}^+(X)$. Here and hence forth $\mathcal{B}^+(X)$ denotes all non–negative Borel functions on X. The above measure μ is called the Revuz measure of (A_t). The set of all Revuz measures of PCAF's is denoted by $\mathcal{S}$. An element of $\mathcal{S}$ is called a smooth measure [F1]. For $\mu \in \mathcal{S}$ we shall denote by (A_t^μ) the unique (up to equivalence) PCAF such that the Revuz measure of (A_t^μ) is μ. A simple analytical description of $\mathcal{S}$ was given in [F1]. A restatement of this description in term of an $\mathcal{E}$–nest is given in Th. 3.3 below.
A Borel measure μ is called of finite energy integral if there exists a constant $C > 0$ such that

$$\int \varphi d\mu \leq C \sqrt{\mathcal{E}_1(\varphi, \varphi)} \quad , \quad \forall \varphi \in \mathcal{F} \cap C_0(X) \, . \tag{3.2}$$

The subset of the measures having a finite energy integral will be denoted denoted by $\mathcal{S}_0$. It is known that $\mathcal{S}_0 \subset \mathcal{S}$ and that each element of $\mathcal{S}$ can be approximated by elements of $\mathcal{S}_0$ ([F1] Th. 3.2.3). A restatement of this approximation procedure will be given in Th. 3.3 below.
In [AM3] we introduced the set $\mathcal{S}_K$, consisting of the smooth measures in Kato class, which is defined by

$$\mathcal{S}_K = \left\{ \mu \in \mathcal{S} : \lim_{t \downarrow 0} \left(\inf_{Cap(N)=0} \sup_{x \in X-N} E_x A_t^\mu \right) = 0 \right\} . \tag{3.3}$$

For a closed set F, let us put

$$\mathcal{F}_F = \{ f \in \mathcal{F} : f = 0 \text{ m.a.e. on } X - F \} \tag{3.4}$$

An increasing sequence of closed sets $\{F_n\}_{n \geq 1}$ is called an $\mathcal{E}$-nest if $\bigcup \mathcal{F}_{F_n}$ is $\mathcal{E}_1$-dense in $\mathcal{F}$.
We denote by σ_B the hitting time of a subset $B \subset X$, i.e. $\sigma_B = \inf\{t > 0 : X_t \in B\}$.

3.1. Lemma Let $\{F_n\}_{n \geq 1}$ be an increasing sequence of closed sets. Then the following three statements are all equivalent.
(i) $\{F_n\}_{n \geq 1}$ is an $\mathcal{E}$-nest.
(ii) $\lim_{n \to \infty} Cap(K - F_n) = 0$, for all compact sets K.
(iii) $P_x(\lim_{n \to \infty} \sigma_{X - F_n} \geq \zeta) = 1$, q.e. $x \in X$.

Proof (ii) $\Leftrightarrow$ (iii) was proved in [F1] Lemma 5.1.6. We now prove (i) $\Leftrightarrow$ (iii). We put

$$\mathcal{F}_0 = \left\{ u : u = E_\cdot \left[\int_0^{\sigma_{X - F_n}} e^{-t} f(X_t) dt \right] \quad , \quad \forall f \in L^2(X, m), n \geq 1 \right\} .$$

Then $\mathcal{F}_0 \subset \bigcup \mathcal{F}_{F_n}$ (see e.g. [FO]). If (iii) holds, then one can show that for each $f \in L^2(X, m)$, the resolvent $R_1 f$ can be approximated by elements of $\mathcal{F}_0$ in $\mathcal{E}_1$-norm, and hence $\mathcal{F}_0$ is $\mathcal{E}_1$-dense in $\mathcal{F}$, proving (i). For proving (i) $\Rightarrow$ (iii), we take a strictly positive function $f \in L^2(X, m)$ and put

$$h_n = R_1 f - E_\cdot \left[\int_0^{\sigma_{X - F_n}} f(X_s) ds \right] \ , \ n \geq 1 \ ,$$
$$h_\infty = \lim_{n \to \infty} h_n \ .$$

One can show that $h_\infty \in \mathcal{F}$ and h_∞ is $\mathcal{E}_1$-orthogonal to $\bigcup \mathcal{F}_{F_n}$, moreover h_∞ is quasicontinuous. If (i) is true, then from the above fact we conclude that $h_\infty = 0$ q.e., which in turn implies (iii). ∎

3.2 Remark The concept of $\mathcal{E}$–nest was introduced in [AM1] where the authors found a one to one correspondence between a broader family of Dirichlet forms (including regular Dirichlet forms, non–regular Dirichlet forms, Dirichlet forms over infinite dimensional spaces) and the family of m–perfect processes, i.e., all the normal strong Markov processes having cadlag sample paths up to life times and having regular resolvent functions (including all the symmetric Hunt processes).
We now restate the descriptions of smooth measures in terms of $\mathcal{E}$–nest in the following:

3.3 Theorem Let μ be a Borel measure on X. Then the following assertions are all equivalent:

(i) $\mu \in \mathcal{S}$.
(ii) μ charges no set of zero capacity and there exists an $\mathcal{E}$–nest $\{F_n\}_{n\geq 1}$ consisting of compact sets F_n such that $\mu(F_n) < \infty, \forall n \geq 1$ and $\mu(X - \bigcup F_n) = 0$.
(iii) There exists an $\mathcal{E}$–nest $\{F_n\}_{n\geq 1}$ such that $I_{F_n} \cdot \mu \in \mathcal{S}_0, \forall n \geq 1$ and $\mu(X - \bigcup F_n) = 0$.
(iv) There exists an $\mathcal{E}$–nest $\{F_n\}_{n\geq 1}$ such that $I_{F_n} \cdot \mu \in \mathcal{S}_K \cap \mathcal{S}_0, \forall n \geq 1$ and $\mu(X - \bigcup F_n) = 0$.
(v) There exists an $\mathcal{E}$–nest $\{F_n\}_{n\geq 1}$ consisting of compact sets F_n such that $I_{F_n} \cdot \mu \in \mathcal{S}_{K,0}$ and $\mu(X - \bigcup F_n) = 0$, where

$$\mathcal{S}_{K,0} = \{\mu \in \mathcal{S}_K : \mu(X) < \infty\} \ . \tag{3.5}$$

Proof For the proof of (i) $\Leftrightarrow$ (ii) $\Leftrightarrow$ (iii) see [F1] Th. 3.2.3 and Th. 5.1.3. The implication (iii) $\Rightarrow$ (iv) is trivial. In Lemma 3.3 below we shall show that $\mathcal{S}_{K,0} \subset \mathcal{S}_0$, hence (iv) $\Rightarrow$ (v) is also trivial. (v) $\Leftrightarrow$ (i) is just a restatement of [AM3] Th. 2.4.

3.4 Remark From the above description we see that the family $\mathcal{S}$ is quite large. It contains all Radon measures charging no set of zero capacity. In particular, if $f \in L^1_{\text{loc}}$ then $|f| \cdot m \in \mathcal{S}$. Moreover, there are many smooth measures which are not Radon measures. If each single–point set of X is a set of zero capacity (as in the case of $X = \mathbb{R}^d$, $d \geq 2$ and capacity associated with Brownian motion) then there exists even smooth measures μ's which are nowhere Radon in the sense that $\mu(G) = \infty$ for all non–empty open set $G \subset X$. See [AM3] for more details. The discovery of nowhere Radon smooth measures draws attention to the existence of Schrödinger operators which are singular on each neighbourhood of every point. We shall not discuss their properties here (see however 4.4 below), for details see [AM4]. We now consider in more details the smooth measures belonging to Kato's class.
For a real valued function f on X, we put

$$\|f\|_q = \inf_{Cap(N)=0} \sup_{x\in X-N} |f(x)| \tag{3.6}$$,

and for $\mu \in \mathcal{S}$, we write

$$\mathcal{U}'\mu(x) = E_x\left[\int_0^\infty e^{-t} dA_t^\mu\right] \tag{3.7}$$

<u>3.5 Lemma</u> If $\mu \in \mathcal{S}_K$, then $\|\mathcal{U}'\mu\|_q < \infty$.

<u>Proof</u> By (3.3) we may find a small positive number t such that $\left\| E_\cdot \left[\int_0^t e^{-s} dA_s^\mu \right] \right\|_q \leq 1$. Hence by the additivity property,

$$\begin{aligned}\mathcal{U}'\mu(x) &= E_x\left[\sum_{n\geq 0}\int_{nt}^{(n+1)t} e^{-s}dA_s^\mu\right] \\ &= E_x\left\{\sum_{n\geq 0} e^{-nt}E_{X_{nt}}\left[\int_0^t e^{-s}dA_s^\mu\right]\right\} \leq \frac{1}{1-e^{-t}}\ , \quad \text{q.e. } x \in X\ .\end{aligned}$$

■

<u>3.6 Lemma</u> Let $\mu_1, \mu_2, \nu_1, \nu_2$ be smooth measures and let f_1 and f_2 be non-negative Borel functions. Then

$$E_{f_2\cdot\mu_2}\left[\int_0^t e^{-A_s^{\nu_1}+A_s^{\nu_2}} f_1(X_s)dA_s^{\mu_1}\right] = E_{f_1\cdot\mu_1}\left[\int_0^t e^{-A_s^{\nu_1}+A_s^{\nu_2}} f_2(X_s)dA_s^{\mu_2}\right]$$

for all $0 \leq t \leq \infty$.
For the proof of Lemma 3.6 see [AM4] Th. 3.2. ■

<u>3.7 Lemma</u> If $\mu \in \mathcal{S}_K$ is such that $\mu(X) < \infty$, then $\mu \in \mathcal{S}_0$.

<u>Proof</u> Let us set $f = \mathcal{U}'\mu$. Applying the above lemmas, we have

$$\int_X f^2 dm = E_{f\cdot m}\left[\int_0^\infty e^{-t}dA_t^\mu\right] = E_\mu\left[\int_0^\infty e^{-t}f(X_t)dt\right] \leq \|f\|_q \mu(X) < \infty$$

and

$$\begin{aligned}\lim_{t\downarrow 0}\frac{1}{t}(f-e^{-t}P_t f, f) &= \lim_{t\downarrow 0}\frac{1}{t}E_{f\cdot m}\left[\int_0^t e^{-s}dA_s^\mu\right] \\ &= \lim_{t\downarrow 0}\frac{1}{t}E_\mu\left[\int_0^t e^{-s}f(X_s)ds\right] \leq \|f\|_q\mu(X) < \infty\ .\end{aligned}$$

Consequently $f := \mathcal{U}'\mu \in \mathcal{F}$, which implies $\mu \in \mathcal{S}_0$. ■
Let us put

$$A_{cl}^+ = \{(A_t) \in A_c^+ :\ \text{The exceptional set of } (A_t) \text{ is empty}\} \tag{3.8}$$

and let $\mathcal{S}_1$ be the set of all Revuz measures of A_{cl}^+. We make the convention that for $\mu \in \mathcal{S}_1$ the PCAF (A_t^μ) always denotes an element in A_{cl}^+ among the equivalent PCAF's associated with μ.

3.8 Proposition Suppose that (2.1) holds. Let μ be a Radon measure. If $\mu \in \mathcal{S}_K$, then $\mu \in \mathcal{S}_1$.

Proof Let $\{G_n\}_{n\geq 1}$ be a sequence of relatively compact open sets such that $\bar{G}_n \subset G_{n+1}$ and $G_n \uparrow X$. Set $E_n = \bar{G}_n$. Obviously we have $P_x(\lim_{n\to\infty} \sigma_{X-E_n} \geq \zeta) = 1, \forall x \in X$. By Lemma 3.5 and Lemma 3.7, we have $I_{E_n} \cdot \mu \in \mathcal{S}_{00}$ for each $n \geq 1$, where

$$\mathcal{S}_{00} \equiv \{\mu \in \mathcal{S}_0 : \mathcal{U}'\mu \text{ is essentially bounded and } \mu(X) < \infty\} .$$

Thus by [F4] Th. 3.1 $\mu \in \mathcal{S}_1$. ∎

3.9 Theorem Suppose that Assumption 2.1 is satisfied. Let μ be a Radon measure. Then $\mu \in \mathcal{S}_K$ if and only if

$$\lim_{t\downarrow 0} \sup_{x\in X} \int_0^t \int_X p(s,x,y)\mu(dy)ds = 0 . \tag{3.9}$$

Furthermore, if $\mu \in \mathcal{S}_K$, then the following assertions hold.

(i) $\lim_{t\downarrow 0} \sup_{x\in X} E_x A_t^\mu = 0$ (3.10)

(ii) $E_\cdot A_t^\mu \in C_b(X)$ for each fixed $t > 0$ (3.11)

(iii) $E_\cdot \int_0^\infty e^{-\alpha t} dA_t^\mu \in C_b(X)$ for each fixed $\alpha > 0$ (3.12)

Proof We first prove (3.10) – (3.12). Let $\mu \in \mathcal{S}_K$ and μ be a Radon measure. Then by the above proposition $(A_t^\mu) \in A_{cl}^+$. Let $t > 0$ and write $f_s(x) = E_x[A_t^\mu - A_s^\mu]$ for $0 \leq s \leq t$. Since $f_s = P_s(E_\cdot A_{t-s}^\mu)$, by Assumption 2.1 we see that $f_s \in C_b(X)$, from which it follows that f_s converges to some continuous function, say f_0, with respect to the supremum norm, when $s \downarrow 0$. But obviously we have $\lim_{s\downarrow 0} f_s(x) = E_x A_t^\mu$. Thus (3.11) is proved. Now (3.10) follows from (3.11) and (3.3), and (3.12) is a consequence of (3.10) and of the fact that

$$E_\cdot \int_t^\infty e^{-\alpha s} dA_s^\mu = e^{-t} P_t E_\cdot \int_0^\infty e^{-\alpha s} dA_s^\mu \quad \in C_b(X) .$$

For the "only if" part of (3.9), we observe that Lemma 3.6 implies that

$$E_x A_t^\mu = \int_0^t \int p(s,x,y)\mu(dy)ds \quad m \text{ a.e. } x \in X$$

and consequently

$$\begin{aligned} E_x(A_t^\mu - A_s^\mu) &= P_s(E_\cdot A_{t-s}^\mu) = P_s\left(\int_0^{t-s} \int p(r,\cdot,y)\mu(dy)dr\right)(x) \\ &= \int_s^t \int p(r,x,y)\mu(dy)dr . \end{aligned}$$

Letting $s \downarrow 0$ we obtain from the above formula

$$E_x A_t^\mu = \int_0^t \int_X p(r,x,y)\mu(dy)dr \quad , \quad \forall x \in X,\ t > 0\,, \tag{3.13}$$

which implies (3.9) by virtue of (3.10). For the "if" part, we assume first that $\mu(X) < \infty$ and that (3.9) holds.
Let $f(x) = \int_0^\infty e^{-t} \int_X p(t,x,y)\mu(dy)dt$. Then

$$\sup_x f(x) := M < \infty$$

Consequently

$$\begin{aligned}\int_X f^2 m(dx) \le M \int_X f m(dx) &= M \int \left[\int_0^\infty e^{-t} \int p(t,x,y)m(dx)dt\right] \mu(dy)\\ &\le M\mu(X) < \infty\,.\end{aligned}$$

Similary

$$\lim_{t\downarrow 0} \frac{1}{t}(f - e^{-t}P_t f, f) \le M \lim_{t\downarrow 0} \frac{1}{t} \int \int_0^t e^{-s} \int p(s,x,y)\mu(dy)m(dx) \le M\mu(X)\,.$$

The above shows that $f \in \mathcal{F}$ which implies $\mu \in \mathcal{S}_0$. In turn this implies that if μ is a Radon measure satisfying (3.9) then $\mu \in \mathcal{S}_1$. Repeating the argument of the first part of this proof we obtain again (3.13). Consequently from (3.9) we conclude that $\mu \in \mathcal{S}_K$, which completes the proof. ∎

<u>*3.10 Remark*</u> Let $(\mathcal{E},\mathcal{F})$ be the classical Dirichlet form generated by Brownian motion (X_t) on $\mathbb{R}^d$, then (3.9) shows that a Radon measure μ is in $\mathcal{S}_K$ if and only if

$$\lim_{\alpha\downarrow 0} \sup_{x\in\mathbb{R}^d} \int_{|x-y|<\alpha} \frac{\mu(dy)}{|x-y|^{d-2}} = 0 \quad , \quad d \ge 3$$

$$\lim_{\alpha\downarrow 0} \sup_{x\in\mathbb{R}^d} \int_{|x-y|<\alpha} \left(\log|x-y|^{-1}\right)\mu(dy) = 0 \quad , \quad d = 2$$

$$\sup_x \int_{|x-y|\le 1} \mu(dy) < \infty \quad , \quad d = 1$$

(c.f. [ASi]). In this case $\mathcal{S}_K$ coincides with the generalized Kato class GK_d introduced in [BM1], and a function $f \in L^1_{loc}$ is in Kato class of functions in the sense of [AS1] if and only if $|f| \cdot m \in \mathcal{S}_K$.

4. Feynman-Kac semigroups in L^2 spaces

For a signed Borel measure $\mu = \mu^+ - \mu^-$, it is suggestive to write $\mu \in \mathcal{S} - \mathcal{S}$ if $\mu^+ \in \mathcal{S}$ and $\mu^- \in \mathcal{S}$. The notations $\mathcal{S} - \mathcal{S}_K, \mathcal{S}_K - \mathcal{S}_K$ are understood in a similar way. For $\mu \in \mathcal{S} - \mathcal{S}$, we write $A_t^\mu = A_t^{\mu^+} - A_t^{\mu^-}$, and refer μ as the Revuz measure of (A_t^μ). Let us introduce the notations, for $\mu \in \mathcal{S} - \mathcal{S}$:

$$P_t^\mu f(x) = E_x\left[e^{-A_t^\mu} f(X_t)\right] , \tag{4.1}$$

$$R_\alpha^\mu f(x) = E_x\left[\int_0^\infty e^{-\alpha t} f(X_t) dt\right] , \tag{4.2}$$

provided the right hand sides make sense. Notice that $(P_t^\mu)_{t>0}$ is the so called Feynman–Kac functional. In the case where (X_t) is the Brownian motion on $\mathbb{R}^d$, $\mu(dx) = V(x)m(dx)$, m being here Lebesgue measure, with $V_- \in K_d$ and $V_+ \in K_d^{loc}$, it is known that $(P_t^\mu)_{t>0}$ is a realization of the Schrödinger semigroup. More precisely, one knows that

$$e^{tH^\mu} f = E_\cdot\left[e^{-\int_0^t V(X_s)ds} f(X_t)\right] \tag{4.3}$$

with $H^\mu = \frac{\Delta}{2} - \mu$. See [Sim1] and references therein for a detailed discussion of this connection. There are also results providing more general conditions on V for the validity of (4.3). See [J] for an analytical approach of recent development on this subject. In what follows we state a general result giving necessary and sufficient conditions for $(P_t^\mu)_{t>0}$ to be a strongly continuous semigroup on $L^2(X, m)$, and discuss its connection with the operator $H^\mu := H - \mu$, where H is the L^2–generator of the underlying process (X_t). This result was obtained in our recent work [AM4] concerning a general perturbation theory of Dirichlet forms.

From now on let us use the short notation $L^2(\mu)$ for $L^2(X, \mu)$. For $\mu \in \mathcal{S} - \mathcal{S}$, we set

$$Q_\mu(f, g) = \int_X f(x)g(x)\mu(dx) \quad , \quad \forall f, g \in L^2(|\mu| + m) \tag{4.4}$$

By Theorem 3.3 (ii) one can show that $L^2(|\mu| + m)$ is dense in $L^2(m)$. Hence Q_μ is a quadratic form on $L^2(m)$. We put

$$\mathcal{E}^\mu(f, g) = \mathcal{E}(f, g) + Q_\mu(f, g) \quad , \quad \forall f, g \in \mathcal{F}^\mu \tag{4.5}$$

where

$$\mathcal{F}^\mu = \mathcal{F} \cap L^2(|\mu| + m) . \tag{4.6}$$

4.1 Theorem Let $\mu \in \mathcal{S} - \mathcal{S}$. Then each of the following conditions (i) – (iv) is a necessary and sufficient condition for the Feynman–Kac functional $(P_t^\mu)_{t>0}$ to be a strongly continuous semigroup on $L^2(m)$:

(i) There exists a constant $\alpha > 0$ such that

$$R_\alpha^\mu f \in L^2(m) \quad , \quad \forall f \in L^2(m)$$

(ii) There exist positive constants c and β such that

$$\|P_t^\mu f\|_{L^2(m)} \leq c e^{\beta t} \|f\|_{L^2(m)} \quad , \quad \forall f \in L^2(m)$$

(iii) Q_{μ^-} is relatively form bounded with respect to $(\mathcal{E}^{\mu^+}, \mathcal{F}^{\mu^+})$ with bound ≤ 1.
(iv) $(\mathcal{E}^\mu, \mathcal{F}^\mu)$ is lower semibounded.

Moreover, if one of the above conditions holds, then P_t^μ is a bounded symmetric operator on $L^2(m)$ for each $t > 0$, and the generator H^μ of (P_t^μ) is a realization of $(H - \mu)$ where H is the L^2–generator of (X_t), in the sense that the closed lower semibounded quadratic form generated by (P_t^μ) is a weak closure of the form $(\mathcal{E}^\mu, \mathcal{F}^\mu)$.

Proof For a verification of conditions (i) – (iv) see [AM4] Th. 4.1. For the final assertion see [AM4] Th. 5.5. ∎

We shall always denote by H^μ the L^2–generator of $(P_t^\mu)_{t>0}$ and by $D(H^\mu)$ the domain of H^μ.

4.2 Theorem A necessary and sufficient condition in order that $(P_t^\mu)_{t>0}$ be a strongly continuous symmetric semigroup on $L^2(m)$ and that $D(H^\mu) \subset \mathcal{F}^\mu$, is that there exists $\alpha > 0$ such that

$$R_\alpha^\mu f \in L^2(m + \mu^-) \quad , \quad \forall f \in L^2(m) \tag{4.7}$$

Proof See [AM4] Theorem 4.5. ∎

Remark Theorem 4.2 gives in fact a necessary and sufficient condition for $(\mathcal{E}^\mu, \mathcal{F}^\mu)$ to be bounded below and closable, with $D(H^\mu) \subset \mathcal{F}^\mu$. This result can be viewed as an improvement of KLMN Theorem in general framework of the theory of perturbations of Dirichlet forms. See [AM4] for more details.
In the case where $X = \mathbb{R}^d$, M is the Lebesgue measure on $\mathbb{R}^d$, and H is a symmetric differential operator on $\mathbb{R}^d$ (as in Example 2.7), we have an explicit description of H^μ, given in the following:

4.3 Theorem Let $X = \mathbb{R}^d$, m be the Lebesgue measure on $\mathbb{R}^d$, and the generator H of (X_t) be a symmetric differential operator on $\mathbb{R}^d$. Suppose that (4.7) holds true, and suppose that μ^- (or μ^+) is a Radon measure. Then

$$D(H^\mu) \subset \mathcal{F}^\mu \cap L^1_{loc}(|\mu|) \tag{4.8}$$

and

$$H^\mu u = (H - \mu)u \quad \text{in distributional sense, i.e.,}$$

$$(u, H\varphi) - \langle u, \varphi\rangle_\mu = (H^\mu u, \varphi) \quad , \quad \forall u \in D(H^\mu), \varphi \in C_0^\infty(\mathbb{R}^d)\ . \tag{4.9}$$

Proof Let $\alpha > 0$ satisfying (4.7). It is well known that

$$D(H^\mu) = \{R^\mu_\alpha f : f \in L^2(m)\} \tag{4.11}$$

and

$$H^\mu(R^\mu_\alpha f) = \alpha(R^\mu_\alpha f) - f \tag{4.12}$$

Let $u = R^\mu_\alpha f$ for some $f \in L^2(m)$. Then $u \in \mathcal{F}^\mu$ and without less of generality we assume that $f \geq 0$. By the generalized resolvent formula (see [AM3] Th. 3.1) we may write

$$u(x) = R_\alpha f - E_x \int_0^\infty e^{-\alpha t} u(X_t) dA_t^{\mu^+} + E_x \int_0^\infty e^{-\alpha t} u(X_t) dA_t^{\mu^-} \tag{4.13}$$

We first assume that f is bounded and $\mu^-(\mathbb{R}^d) < \infty$. Then the three terms on the left hand side of (4.13) are all in $\mathcal{F}$. Notice that any function $\varphi \in C_0^\infty(\mathbb{R}^d)$ can be expressed as the difference of two α–excessive functions in $\mathcal{F}$. We obtain from the Revuz formula (3.1):

$$(u, -H\varphi) + \alpha(u, \varphi) + \langle u, \varphi\rangle_{\mu^+} = (f, \varphi) + \langle u, \varphi\rangle_{\mu^-} \ , \ \forall \varphi \in C_0^\infty(\mathbb{R}^d) \tag{4.14}$$

(4.14) holds in fact without assuming that f is bounded and that μ^- is finite, since in the general case we may find bounded $f_n \uparrow f$ and finite ${\mu^-}_n \uparrow \mu^-$, and take the limit $n \to \infty$ in both sides of (4.14). (Notice that $u \in L^2(\mu^-) \subset L^1_{loc}(\mu^-)$ because μ^- is a Radon measure.) In particular (4.14) implies $u \in \mathcal{F}^\mu \cap L^1_{loc}(|\mu|)$. Comparing (4.14), (4.11) and (4.12) we obtain the lemma. ∎

We shall now consider the particular case where $H = \frac{\Delta}{2}$, i.e. (X_t) being the Brownian motion on $\mathbb{R}^d$. Then Theorem 4.3 shows that H^μ is the Schrödinger operator $\frac{\Delta}{2} - \mu$. Notice that when $d \geq 2$ there exist nowhere Radon smooth measure μ's. This will produce Schrödinger operators which are singular on each neighbourhood of every point $x \in \mathbb{R}^d$. A concrete example of such singular Schrödinger operators is given in the following proposition.

4.4 Proposition Let $\{x_i\}_{i\geq 1}$ be a dense subset of $\mathbb{R}^d$, $d \geq 2$, and $\{\alpha_i\}_{i\geq 1}$ be a sequence of real numbers such that $\alpha_i \geq d, \forall i \geq 1$. There exists then a sequence of strictly positive numbers $\{c_i\}_{i\geq 1}$ such that the measure $V \cdot m$ with V being defined by

$$V(x) = \sum_{i\geq 1} c_i |x - x_i|^{-\alpha_i}$$

is a smooth measure (w.r.t. the classical Dirichlet form).

For the proof of Proposition 4.4 see [AM3]. ■

Applying Theorem 4.3 to $V \cdot m$ we see that there exists a singular Schrödinger operator H^μ defined on a dense subset $D(H^\mu)$ of $L^2(\mathbb{R}^d, m)$, such that $D(H^\mu) \subset \mathcal{F} \cap L^2(V \cdot m)$ and

$$H^\mu u = (\frac{\Delta}{2} - V)u \quad , \quad \forall u \in D(H^\mu)$$

in the distributional sense.

5.Compatibility of smooth measures

For $\mu, \nu \in \mathcal{S}$ and $\alpha \geq 0$, we define the operator $G_\nu^{\mu+\alpha}$ by

$$G_\nu^{\mu+\alpha} f(x) = E_\nu \left[\int_0^\infty e^{-A_t^\mu - \alpha t} f(X_t) dA_t^\mu \right] \tag{5.1}$$

provided the right hand side makes sense.

5.1 Theorem Let $\mu, \nu \in \mathcal{S}$. Then the following statements are all equivalent:

(i) $\inf_{\alpha>0} \|G_\nu^{\mu+\alpha} 1\|_q < \infty$ and $\inf_{\alpha>0, n\geq 1} \|(G_\nu^{\mu+\alpha})^n 1\|_q < 1$ (5.2)
(1 stands for the constant function with value 1)

(ii) $\|E.\left[\int_0^\infty e^{-A_s^\mu + A_s^\nu - \alpha s} dA_s^\nu\right]\|_q < \infty$ for some $\alpha > 0$ (5.3)

(iii) $\|E.\left[\int_0^t e^{-A_s^\mu + A_s^\nu} dA_s^\nu\right]\|_q < \infty$ for some $t > 0$ (5.4)

(iv) $\|E.\left[e^{-A_t^\mu + A_t^\nu} + \int_0^t e^{-A_s^\mu + A_s^\nu} dA_s^\mu\right]\|_q < \infty$ for some $t > 0$ (5.5)

(v) There exists a set N with $Cap(N) = 0$ such that for all $0 < T < \infty$,

$$\sup_{t\leq T} \sup_{x\in X-N} E_x \left[e^{-A_t^\mu + A_t^\nu} + \int_0^t e^{-A_s^\mu + A_s^\nu} d(A_s^\mu + A_s^\nu) \right] < \infty \,. \tag{5.6}$$

<u>*Proof*</u> (ii) ⇔ (iii) is trivial. (iii) ⇔ (iv) follows from the fact that

$$1+\int_0^t e^{-A_s^\mu+A_s^\nu}dA_s^\nu = e^{-A_s^\mu+A_s^\nu}+\int_0^t e^{-A_s^\mu+A_s^\nu}dA_s^\mu \ . \tag{5.7}$$

By (ii) we may find a set N with $Cap(N)=0$ such that

$$\sup_{x\in X-N}\left|E_x\left[\int_0^\infty e^{-A_s^\mu+A_s^\nu-\alpha s}dA_s^\nu\right]\right|<\infty \ . \tag{5.8}$$

Thus (5.6) follows from (5.8) and (5.7) which proves that (ii) ⇒ (v). (v) ⇒ (iii) is trivial. It remains to prove (i) ⇔ (ii). To this end we remark first that

$$(G_\nu^{\mu+\alpha}1)^{n+1}1(x)=E_x\left[\frac{1}{n!}\int_0^\infty e^{-A_s^\mu-\alpha s}(A_s^\nu)^n dA_s^\nu\right] \tag{5.9}$$

Indeed, (5.9) is true for $n=1$. Suppose that (5.9) is true for $n=k$, then we have for q.e. $x\in X$,

$$\begin{aligned}(G_\nu^{\mu+\alpha})^{k+1}1(x)&=E_x\left\{\int_0^\infty e^{-A_t^\mu-\alpha t}\left[E_{X_t}\frac{1}{(k-1)!}\int_0^\infty e^{-A_s^\mu-\alpha s}(A_s^\nu)^{k-1}dA_s^\nu\right]dA_t^\nu\right\}\\&=E_x\left\{\int_0^\infty e^{-A_t^\mu-\alpha t}\left[\int_t^\infty \frac{1}{(k-1)!}e^{-A_s^\mu-\alpha s}(A_s-A_t)^{k-1}dA_s^\nu\right]dA_t^\nu\right\}\\&=E_x\left[\frac{1}{k!}\int_0^\infty e^{-A_s^\mu-\alpha s}(A_s^\nu)^k dA_s^\nu\right] \ .\end{aligned}$$

Thus by induction (5.9) is true for any $n\geq 1$. Now (i) ⇒ (ii) follows easily from (5.9). We shall now prove the implication (i) → (ii), so we assume that (ii) holds. Then $T:=G_\nu^{\mu+\alpha}$ is a linear bounded positive operator from $\mathcal{B}_{qb}$ to $\mathcal{B}_{qb}$, where $\mathcal{B}_{qb}$ is the Banach space of all quasi–bounded Borel functions on X equipped with the norm $\|\cdot\|_q$. Set $\lambda\equiv\lim_{n\to\infty}\|T^n\|^{\frac{1}{n}}$. If $\lambda\geq 1$, then by (5.9) and (5.3) one would have

$$\|\sum_{n=0}^\infty(\frac{T}{\lambda})^n 1\|_q\leq\|\sum_{n=0}^\infty T^n 1\|_q<\infty \ ,$$

which would imply that λ is in the resolvent set $\rho(T)$ of T. But it is well known that λ can not be in $\rho(T)$ (c.f. e.g. [Sch] pp. 263). Consequently we must have $\lambda<1$, which implies (i). ■

For a given $\mu\in\mathcal{S}$, we shall say that a smooth measure ν is <u>compatible</u> with μ, if ν satisfies one of the conditions in Th. 5.1. The set of all smooth measures which are compatible with μ will be denoted by $C(\mu)$. The following Proposition holds:

5.2 Proposition For any $\mu \in \mathcal{S}$, we have $\mathcal{S}_K \subset C(\mu)$.

Proof Let O be the vanishing measure on X. Let $\nu \in \mathcal{S}_K$, i.e. ν satisfies (3.3), than one can obtain from (3.3) that

$$\inf_{\alpha>0} \|G_\nu^{O+\alpha}1\|_q = 0 \tag{5.10}$$

Thus by (5.2) we have $\nu \in C(O) \subset C(\mu)$.

5.3 Remark Let $X = \mathbb{R}^d$ and m be the Lebesgue measure.
We put for $V \in L^1_{loc}(\mathbb{R}^d, m)$,

$$K_d(V) = \inf_{\alpha \downarrow 0}\left(\text{ess} \sup_{x\in\mathbb{R}^d} \int_{|x-y|\leq\alpha} |g_d(x-y)|\,|V(y)|\,dy\right) , \tag{5.11}$$

where $g_1(x) := |x|$, $g_2(x) := \pi^{-1}\log|x|$ and

$$g_d(x) := \frac{\Gamma(\frac{d}{2})}{(d-2)}\pi^{-\frac{d}{2}}|x|^{-\frac{d}{2}} \quad \text{for } d \geq 3 .$$

Let $(\mathcal{E},\mathcal{F})$ be the classical Dirichlet form on $\mathbb{R}^d$ (see Example 2.6). It is easy to see that $K_d(V) < 1$ implies $|V|\cdot m \in C(0) \subset C(\mu)$ for any $\mu \in \mathcal{S}$. On the other hand, V is in Kato class, i.e., $|V|\cdot m \in \mathcal{S}_K$, if and only if $K_d(V) = 0$. An example of a function V which does not belong to Kato's class but satisfies $K_d(V) < 1$ was given in [AS] Appendix 1, which illustrates the fact that $\mathcal{S}_K$ is a proper subset of $C(\mu)$.

The following example shows that for some choice of $\mu \in \mathcal{S}$, the class $C(\mu)$ contains even function V with $K_d(V) = \infty$.

5.4 Example Consider again the classical Dirichlet form on $\mathbb{R}^d, d \geq 2$. We define for some $k \geq 2$:

$$V^+(x) = |x|^{-k}\left(\frac{1}{2} - \sin(|x|^{-k})\right) \vee 0 ,$$

$$V^-(x) = |x|^{-k}\left(\sin(|x|^{-k}) - \frac{1}{2}\right) \vee 0 ,$$

and set $\mu = V^+ \cdot m$, $\nu = V^- \cdot m$. We have then $K_d(V^+) = \infty$ and $K_d(V^-) = \infty$. But it can be checked that both μ and ν are smooth measures and $\nu \in C(\mu)$. This example was found by Sturm. For details see [St].

6. L^p–boundedness and L^p–smoothing

This section is devoted to the proof of the following Theorem.

6.1 Theorem Let $\mu \in \mathcal{S} - \mathcal{S}$ be such that $\mu^- \in C(\mu^+)$. Then we have:
(i) There exist constants c and β such that

$$\|P_t^\mu\|_{p,p} \leq ce^{\beta t} \quad , \quad \forall 1 \leq p \leq \infty \; , \; t > 0 \tag{6.1}$$

(ii) $(P_t^\mu)_{t>0}$ is a strongly continuous symmetric semigroup on $L^2(m)$ and $D(H^\mu) \subset \mathcal{F}^\mu := \mathcal{F} \cap L^2(|\mu|)$. Here H^μ denotes the L^2–generator of $(P_t^\mu)_{t>0}$.
(iii) If Assumption 2.2 holds, then

$$\|P_t^\mu\|_{p,p'} < \infty \quad , \quad \forall 1 \leq p \leq p' \leq \infty \; , \; \forall t > 0 \tag{6.2}$$

(iv) If Assumption (2.3) holds, then (P_t^μ) satisfies (6.2) and in addition the following bound holds for some constant c':

$$\|P_t^\mu\|_{1,\infty} \leq c' t^{-\frac{d}{2}} \quad , \quad \forall 0 < t \leq 1 \tag{6.3}$$

The proof of Theorem 6.1 is split into several steps.

6.2 Lemma Let $\mu^- \in C(\mu^+)$. Then there exists constant c and β such that

$$\|E_\cdot e^{-A_t^\mu}\|_q \leq ce^{\beta t} \tag{6.4}$$

Proof From (5.6) we infer that there exists a constant $\lambda \geq 1$ such that

$$\sup_{o \leq t \leq 1} \|E_\cdot e^{-A_t^\mu}\|_q \leq \lambda < \infty$$

For $t = n + s$, n being a natural number and $0 \leq s \leq 1$, we obtain by induction

$$E_x e^{-A_t^\mu} = E_x \left[e^{-A_n^\mu} E_{X_n} e^{-A_s^\mu} \right] \leq \lambda^{n+1} \quad , \quad \text{q.e. } x \in X \; . \tag{6.5}$$

Putting $c = \lambda$ and $\beta = \log \lambda$ we obtain (6.4) using (6.5).

6.3 Corollary Let $\mu^- \in \mathcal{S}_K$. Then P_t^μ is a symmetric bounded operator on $L^2(m)$ for each $t > 0$.

Proof $\mu^- \in \mathcal{S}_K$ implies that $2\mu^- \in C(2\mu^+)$. Let c and β be constants satisfying $\|E.e^{-2A_t^\mu}\|_q \leq c^2 e^{2\beta t}$. We have for any $f \in L^2(m)$,

$$|P_t^\mu f|^2 \leq (E.e^{-2A_t^\mu})(E.|f|^2) \leq c^2 e^{2\beta t} E.|f|^2 \quad \text{q.e. } x \in X .$$

Hence

$$\|P_t^\mu f\|_{L^2(m)} \leq e^{\beta t} \|f\|_{L^2(m)}$$

and the proof is completed by applying Theorem 4.1. ■We are now in position of proving Theorem 6.1:

Proof of Theorem 6.1 (i) Let c and β be as in Lemma 6.2. By Th. 3.3 we may take an $\mathcal{E}$-nest $\{F_n\}_{n\geq 1}$ consisting of compact sets F_n (hence $m(F_n) < \infty$, $\forall n \geq 1$) such that $\mu^-(X - \cup F_n) = 0$ and $I_{F_n} \cdot \mu^- \in \mathcal{S}_{K,0}$. Let us put

$$\mu_n = \mu^+ - I_{F_n} \cdot \mu^- \tag{6.6}$$

and

$$f_n = (|f| \wedge n)\, I_{F_n} \tag{6.7}$$

for an arbitrary function f on X. For $f \in L^1(m)$, by the symmetric property of $P_t^{\mu_n}$ on $L^2(m)$ (see Corollary 6.3) and (6.4) we have

$$\int (P_t^{\mu_n} f_n) I_{F_n} m(dx) \leq ce^{\beta t} \int f_n m(dx) \leq ce^{\beta t} \|f\|_{L^1(m)} \, .$$

Hence by the monotone convergence theorem we obtain

$$\|P_t^\mu\|_{1,1} \leq ce^{\beta t} \tag{6.8}$$

Notice that (6.8) implies in particular that

$$P_t^\mu I_E = 0 \quad m \text{ a.e.} , \quad \forall E \in \mathcal{B}(X) \text{ with } m(E) = 0 \tag{6.9}$$

From (6.9) and (6.4) we obtain $\|P_t^\mu\|_{\infty,\infty} \leq ce^{\beta t}$.
For $f \in L^p(m)$ with $1 < p < \infty$, we have

$$|P_t^\mu f|^p \leq \left(E.e^{-A_t^\mu}\right)^{p-1} P_t^\mu |f|^p \tag{6.10}$$

Thus $\|P_t^\mu\|_{p,p} \leq ce^{\beta t}$ by (6.10), (6.8) and (6.4).

Proof of Th. 6.1 (ii) From (5.2) we see that $\mu^- \in C(\mu^+)$ implies $\lambda\mu^- \subset C(\mu^+)$ for some $\lambda > 1$. Thus by (i) there exist constants c_1 and β_1 such that $\|P_t^{(\mu^+ - \lambda\mu^-)}\|_{2,2} \leq c_1 e^{\beta_1 t}$, which in turn implies by Th. 4.1 that Q_{μ^-} is relatively form bounded with respect to $(\mathcal{E}^{\mu^+}, \mathcal{F}^{\mu^+})$ with bound $\leq \lambda^{-1} < 1$. Applying now the KLMN Theorem (see e.g. [RS2]) we see that $(\mathcal{E}^\mu, \mathcal{F}^\mu)$ is bounded from below and closed, proving (ii).

Proof of Th. 6.1 (iii) and (iv) Let n be a natural number satisfying $2^{\frac{1}{n}}\mu^- \in C(\mu^+)$. Set $p = 2^{\frac{1}{n}}$ and let q satisfy $\frac{1}{p}+\frac{1}{q}=1$. Suppose that Assumption 2.3 holds, we have for $o < t < 1$ and $f \in L^q(m)$,

$$|P_t^\mu f| \le \left(E.e^{-pA_t^\mu}\right)^{\frac{1}{p}} (P_t|f|^q)^{\frac{1}{q}} \ ,$$

which together with Assumption 2.3 and Lemma 6.2 yields

$$\|P_t^\mu\|_{q,\infty} \le c_1 t^{-\frac{d}{2q}} \quad , \quad \forall 0 < t \le 1 \tag{6.11}$$

for some constant c_1. For $f \in L^1(m)$ and $g \in L^q(m)$, let f_n and g_n be defined in the same way as in (6.7). Then by (6.11)

$$(P_t^\mu f_n, g_n) = (f_n, P_t^\mu g_n) \le c_1 t^{-\frac{d}{2q}} \|g_n\|_{L^q(m)} \|f_n\|_{L^1(m)} \ .$$

Letting n tends to infinity we obtain

$$\|P_t^\mu\|_{1,p} \le c_1 t^{-\frac{d}{2q}} \quad , \quad 0 < t \le 1 \tag{6.12}$$

(6.12) and (6.10) yields for some constant $c_2 > 0$,

$$\|P_t^\mu\|_{p,p^2} \le c_2 t^{-\frac{d}{2qp}}$$

Repeating the argument we obtain (for some constants c_i)

$$\|P_t^\mu\|_{p^i,p^{i+1}} \le c_{i+1} t^{-\frac{d}{2qp^i}} \quad , \quad \forall 0 < t \le 1 \ , \ i \ge 1 \tag{6.13}$$

Therefore we have

$$\begin{aligned} \|P_t^\mu\|_{1,2} = \|(P_{\frac{t}{n}}^\mu)^n\|_{1,2} &\le c_1 \dots c_{n-1} \left(\frac{t}{n}\right)^{-\frac{d}{2q}(1+\frac{1}{p}+\dots\frac{1}{p^{n-1}})} \\ &= k t^{-\frac{d}{4}} \quad , \quad \forall 0 < t \le 1 \end{aligned} \tag{6.14}$$

with $k := c_1 \dots c_{n-1} n^{\frac{d}{2}}$. By a similar argument as in (6.12) we infer from (6.14)

$$\|P_t^\mu\|_{2,\infty} = \|P_t^\mu\|_{1,2} \le k t^{-\frac{d}{4}} \quad , \quad \forall 0 < t \le 1 \tag{6.15}$$

From (6.14) and (6.15) we obtain (6.3). If we impose Assumption 2.2 instead of 2.3, then the same argument yields

$$\|P_t^\mu\|_{1,\infty} < \infty \quad , \quad \forall 0 < t \le 1 \tag{6.16}$$

instead of (6.3). The proof will be completed by showing that (6.16) implies that (6.2). To this end we observe first that (6.16) and (6.8) implies

$$\|P_t^\mu\|_{1,\infty} < \infty \quad , \quad \forall t > 0 \tag{6.17}$$

Applying (6.10) to (6.17) we obtain

$$\|P_t^\mu\|_{p,\infty} < \infty \quad , \quad \forall t > 0 \ , \ 1 < p < \infty \tag{6.19}$$

Using (6.10) again we are able to conclude that

$$\|P_t^\mu\|_{p,qp} < \infty \quad , \quad \forall t > 0 \ , \ 1 < p < \infty \ , \ 1 < q < \infty \ . \tag{6.20}$$

Comparing (6.20), (6.17) and (6.1) we obtain (6.2), which completes the proof of Theorem 6.1. ∎

6.4 Remark The property (6.2) is called L^p–smoothing by Simon [Sim]. There are many papers concerning L^p–smoothing of Schrödinger semigroups, see [Sim] Section B, and the literature quoted there, see also [V]. In [S] Simon assumed that the potential term of the Schrödinger operator is a function V with $V^+ \in K_d^{loc}$ and $V^- \in K_d$ (he also pointed out that V^+ can be more singular without affecting the result). In [V] Voigt extended the result of [Sim] to the case where V^+ is $U_1(\cdot)$–admissible and $K_d(V^-) < 1$ (it is known that the condition $K_d(V^-) < 1$ is weaker than $V^- \in K_d$, see Remark 5.3.) Let us emphasize that Theorem 6.1 (iii) extends the above results in two directions. On the one hand, the potential term is assumed to be a signed smooth measure (which is more general than $L^1_{loc}(m)$, see Remark 3.4) satisfying $\mu^- \in C(\mu^+)$ (which is more general than $K_d(V^-) < 1$, see Example 5.4). On the other hand, our result holds not only for Schrödinger semigroups (see examples 2.7, 2.9, 2.10). We acknowledge that the method used for proving Theorem 6.1 (iii) and (iv) was motivated by the works of [Sim] and [V].

7. Integral kernels

Throughout this section we will always assume that Assumption 2.4, 2.2 and 2.1 hold. The following result is more or less known.

7.1 Theorem Let $\mu \in \mathcal{S}_K - \mathcal{S}_K$. Then there exists an integral kernel $p^\mu(t,x,y)$ which is jointly continuous on $(0,\infty) \times X \times X$ satisfying
(i) $P_t^\mu f(x) = \int p^\mu(t,x,y) f(y) m(dy) \quad , \quad \forall f \in \mathcal{B}^+(X)\ .$ (7.1)
(ii) $P_t^\mu f(x)$ is jointly continuous on $(0,\infty) \times X$, $\forall f \in L^p(m)$, $1 \leq p \leq \infty$

Remark on the proof of Th. 7.1 For the classical case of (X_t) being the Brownian motion on $\mathbb{R}^d$ and $\mu = V \cdot m$ with $V^+ \in K_d^{loc}$, $V^- \in K_d$, a complete proof was given in [Sim]. For general Feller process satisfying Assumptions 2.4, 2.2 and 2.1, and $\mu = V \cdot m$, with V being a function in Kato class, a complete, but rather complicated proof was given in [C].
For $\mu \in \mathcal{S}_K - \mathcal{S}_K$ being signed smooth measure in Kato class, in the case of (X_t) being the Brownian motion on $\mathbb{R}^d$,a proof was given in [BM1]. The general case can be proved along the same line of arguments used in [BM1]. ∎

Notice that in the classical case, (X_t) being the Brownian motion on $R^d, V^+ \in K_d^{\rm loc}$ does not imply that $V^+ \cdot m \in \mathcal{S}_K$. hence Theorem 7.1 can not completely cover the corresponding result given in [Sim]. In the remainder of this section we shall study the integral kernel of $\mathcal{P}_t^\mu$ for the general case of $\mu \in \mathcal{S} - \mathcal{S}$ with $\mu^- \in C(\mu^+)$, which will therefore provide a complete generalization of the corresponding result in [Sim]
Let us define for $\mu \in \mathcal{S}$

$$K(\mu) = \{x \in X : \text{ there exists a neighborhoud } E \text{ of } x \text{ such that } I_E \cdot \mu \in \mathcal{S}_{K,0}\} \quad (7.3)$$

Notice that if $\mu = V \cdot m$ with $V \in K_d^{\rm loc}$, then $K(\mu) = X$.

7.2 Lemma Let $\mu \in \mathcal{S}$. Then there exists an $\mathcal{E}$–nest $\{F_n\}_{n\geq 1}$ satisfying Th. 3.3 (v) such that $K(\mu) \subset \bigcup_{n\geq 1} F_n^0$, where F_n^0 denotes the interior of F_n.

Proof Let $\{E_n\}_{n\geq 1}$ be an $\mathcal{E}$–nest satisfying Th. 3.3 (v), and let $\{G_n\}_{n\geq 1}$ be an increasing sequence of relatively compact open sets such that

$$\bar{G}_n \subset G_{n+1} \quad \text{and} \quad K(\mu) = \bigcup_{n=1}^{\infty} G_n \, .$$

Define $F_n = \bar{G}_n \cup E_n$. Then the sequence $\{F_n\}_{n\geq 1}$ fulfills all conditions claimed in Lemma 7.2.
For an arbitrary subset $B \subset X$, we put

$$\tau_B = \sigma_{X-B} := \inf\{t > 0 : X_t \in X - B\} \tag{7.4}$$

Let $\mu \in \mathcal{S} - \mathcal{S}$. We take an $\mathcal{E}$–nest $\{F_n\}_{n\geq 1}$ satisfying Lemma 7.2 with respect to $|\mu| := \mu^+ + \mu^-$. In what follows we shall always denote by

$$\mu_n^+ = I_{F_n} \cdot \mu^+, \quad \mu_n^- = I_{F_n} \cdot \mu^- \quad \forall n \geq 1 \tag{7.5}$$

and

$$\mu_{n\cdot k} = \mu_n^+ - \mu_k^-, \quad \mu_{\infty,k} = \mu^+ - \mu_k^-, \quad \forall n \geq 1, k \geq 1 \tag{7.6}$$

Using now the one to one correspondence between $\mathcal{S}_1$ and A_{cl}^+ (see [F4]), we have for $n \geq k$,

$$P_x\left(I_{F_k} \cdot A_t^{\mu_n^+} = A_t^{\mu_k^+} \, , \, \forall t \geq 0\right) = 1 \, , \quad \forall x \in X \tag{7.7}$$

where $I_{F_k} \cdot A_t^{\mu_m^+} := \int_0^t I_{F_k}(X_s) dA_s^{\mu_n^+}$. Put

$$\tau_n = \tau_{F_n}, \quad \tau_\infty = \lim_{n\to\infty} \tau_n \tag{7.8}$$

and define

$$A_t^{\mu^+}(w) = \begin{cases} A_t^{\mu_n^+}(w) & , \forall \tau_{n-1}(w) \leq t \leq \tau_n(w) \quad (\tau_0 := 0) \\ \lim_{t\uparrow\tau_\infty} A_t^{\mu_n^+}(w) & , \forall t \geq \tau_\infty(w) \end{cases} \tag{7.9}$$

$$N = \{x \in X : P_x(\tau_\infty \geq \zeta) < 1\} \, . \tag{7.10}$$

By (7.7) and Lemma 3.1 (iii), we have $(A_t^{\mu^+}) \in A_c^+$ with N being an exceptional set of $(A_t^{\mu^+})$. It is easy to check that $(A_t^{\mu^+})$ is one of the equivalent class of the PCAF's corresponding to μ^+. In what follows we will always assume that $A_t^{\mu^+}$ is specified by (7.9) and that $(A_t^{\mu^-})$ is specified in the same way.

<u>7.3 Lemma</u> Let $\mu \in \mathcal{S} - \mathcal{S}$ and τ_n be specified by (7.10). If Assumption 2.6 is true, then for any compact set $F \subset K(\mu)$, we have

$$\lim_{n\to\infty} \sup_{x\in F} P_x\{t \wedge \rho > \tau_n\} = 0, \quad \forall t > 0 \tag{7.11}$$

<u>Proof</u> By Lemma 7.3 we may assume that $F \subset F_k^0$ for some $k \geq 1$. Let us denote τ by

$$\tau = \tau_{F_k^0} := \inf\{t > 0 : X_t \in X - F_k^0\} \tag{7.12}$$

Then $\tau_n \geq \tau$ for all $n \geq k$. Fix now $t > 0$ and define for $x \in F_k^0$,

$$f_n(x) = E_x\left[E_{X_\tau}\left(I_{\{t\wedge\rho>\tau_n\}}\right)\right] \tag{7.13}$$

By Lemma 3.1 (iii) we have $f_n(x) \downarrow 0$ for all $x \in F_k^0$ when $n \to \infty$. By Assumption 2.6 f_n is continuous for each n. Consequently f_n tends to zero uniformly on each compact subset of F_k^0. In particular,

$$\lim_{n\to\infty} \sup_{x\in F} f_n(x) = 0 \ .$$

The claimed assertion now follows from the fact that for $n \geq k$,

$$\begin{aligned} P_x\{t \wedge \rho > \tau_n\} &\leq P_x\{t \wedge \rho > \tau_n \circ \Theta_\tau\} \\ &= E_x\left[E_{X_\tau}\left(I_{\{t\wedge\rho>\tau_n\}}\right)\right] = f_n(x), \quad \forall x \in F_k^0 \ . \end{aligned}$$

<u>7.4 Corollary</u> If Assumption 2.11 is true, then

$$K(\mu) \subset X - N$$

with N specified by (7.10).

<u>Proof</u> It follows immediately from (7.11) and (7.10).
We are now in a position to state the main result of this section.

7.5 Theorem Let $\mu \in \mathcal{S} - \mathcal{S}$ with $\mu^- \in C(\mu^+)$. Then there exists an integral kernel $p^\mu(t,x,y)$ which satisfies the following properties.

(i) $\int p^\mu(t,x,y)f(y)m(dy) = \mathcal{P}_t^\mu f(x) := E_x\left[e^{-A_t^\mu} f(X_t)\right]$,
$\forall \quad f \in \mathcal{B}^+(X), \quad t > 0, \quad x \in X - N$ (7.14)
Here and henceforth N is specified by (7.10).

(ii) $\int p^\mu(t,x,z)p^\mu(s,z,y)m(dz) = p^\mu(t+s,x,y), \quad \forall \quad t,x,y$ (7.15)

(iii) $p^\mu(t,x,y) = p^\mu(t,y,x), \quad \forall \quad t.x.y$ (7.16)

(iv) For each $o < s < T < \infty$, we have
$\{\sup p^\mu(t,x,y) : s \leq t \leq T, x,y \in X - N\} < \infty$ (7.17)

(v) If Assumption 2.11 holds, then
$p^\mu(t,x,y)$ is continuous on $(0,\infty) \times K(|\mu|) \times K(|\mu|)$ (7.18)
and for each $f \in \mathcal{B}_b(X)$,
$\int p^\mu(t,x,y)f(y)m(dy)$ is continous on $(0,\infty) \times (K(\mu)$. (7.19)

Proof If $\mu \in \mathcal{S}_K - \mathcal{S}_K$, then we have a continuous kernel $p^\mu(t,x,y)$ satisfying (7.1). By the semigroup property and the symmetry property we see that $p^\mu(t,x,y)$ satisfies (7.15) and and (7.16) for m-a.e.x and m-a.e.y. Thus by the continuity of $p^\mu(t,x,y)$ and the dominated comvergence theorem if follows that $p^\mu(t,x,y)$ verifies (7.15) and (7.16) for all (t,x,y) in $(0,\infty) \times X \times X$. For the general case in which $\mu \in \mathcal{S} - \mathcal{S}$ with $\mu^- \in C(\mu^+)$, we take an $\mathcal{E}$-nest $\{F_n\}$ satisfying Lemma 7.2. Denote by $q_{n,k}(t,x,y)$ the corresponding continuous integral kernel of $P_t^{\mu_{n,k}}$ where $\mu_{n,k}$ is defined by (7.6). Obviously $q_{n,k}$ decreases pointwise when n increases, and $q_{\infty,k} := \lim_{n\to\infty} q_{n,k}$ increases pointwise when k increases. We now first let $n \to \infty$ and then let $k \to \infty$. Denote the limit function by $p^\mu(t,x,y)$. Then $p^\mu(t,x,y)$ satisfies the properties (i) – (iii) of Th. 7.5 by the monotone convergence theorem. We will now prove that $p^\mu(t,x,y)$ satisfies also (iv) and (v). Given $0 < s < T < \infty$ we put

$$C_1 := \|P^\mu_{\frac{s}{2}}\|_{1,\infty}, \quad C_2 = \sup_{t \leq T} \sup_{x \in X-N} E_x e^{-A_t^\mu}. \tag{7.20}$$

We conclude from (6.18) and (5.6) that C_1 and C_2 are finite. (In (5.6) the exceptional set N is not specified by (7.10). But it is easy to specify it by using an aproximation of μ by the sequence $\{\mu_{n,k}\}$ defined in (7.6).) Taking into account the fact that for m-a.e. y,

$$p^\mu(t,x,y) \leq \|P_t^\mu\|_{1,\infty}, \quad \text{m-a.e.}x$$

We obtain from (7.20) and (7.14)

$$p^\mu(t,x,y) \leq C_1 C_2, \quad \text{m-a.e.}y, \quad \forall \quad \frac{s}{2} \leq T, \quad x \in X - N$$

By the symmetry property (7.16), the above inequality implies

$$p^{\mu}(t,x,y) \leq C_1C_2, \quad \text{m-a.e.}x, \quad \forall \quad \frac{s}{2} < t \leq T, \quad y \in X - N$$

Applying (7.14) again we obtain (iv).
We now prove (v) under the additional Assumption 2.11. Let $f \in \mathcal{B}_b(X)$. Without loss of generality we asume that $\|f\|_{L^\infty} \leq 1$.
We define

$$g_{n,k}(t,x) = \int p^{\mu_{n,k}}(t,x,y)f(y)m(dy), \quad 1 \leq n \leq \infty, k \geq 1 \tag{7.21}$$

$$g(t,x) = \int p^{\mu}(t,x,y)f(y)m(dy). \tag{7.22}$$

For fixed $\tau < 0$ and $k \geq 1$, we have for $x \in X - N$ and $0 < t \leq T$,

$$|g_{\infty,k}(t,x) - g_{n,k}(t,x)| \leq E_x\left[2e^{A_t^{\mu_n^-}} I_{\{t\wedge\rho>\tau_n\}}\right]$$

$$\leq 2(E_xe^{2A_T^{\mu_n^+}})^{\frac{1}{2}}(P_x\{T\wedge\rho>\tau_n\})^{\frac{1}{2}} \tag{7.23}$$

Applying Lemma 7.3 we conclude from (7.23) that for any compact set $F \subset K(|\mu|)$,

$$\lim_{n\to\infty}\{sup|g_{\infty,k}(t,x) - g_{n,k}(t,x)| \quad : \quad t\in[0,T], x \in F\} = 0.$$

Hence $g_{\infty.k}$ continuous on $(0,\infty)\times K(|\mu|)$ owing to the fact that $\{g_{n,k}\}$ are continuous functions and that $T > 0$ and $F \subset K(|\mu|)$ have been arbitrarily choosen. Similary we have for $T > 0$,

$$|g(t,x) - g_{\infty,k}(t,x)| \leq E_x\left[2e^{-A_t^{\mu}} I_{\{t\wedge\rho>\tau_n\}}\right)]$$

$$2(E_xe^{-pA_T^{\mu}})^{\frac{1}{p}}(P_x\{T\wedge\rho>\tau_k\})^{\frac{1}{q}} \tag{7.24}$$

where $\frac{1}{p}+\frac{1}{q}=1$ and $p>1$ has to be choosen in such a way that $p\mu^- \in C(p\mu^+)$. Applying Lemma 7.3 again we see that g is continuous on $(o,\infty)\times K(|\mu|)$, proving (7.19). For fixed $y \in X-N$ and $s > 0$, we apply (7.19) to the function $f(x) := p^{\mu}(s,x,y)$. By write of (7.17) and (7.15) we see that for fixed $y \in X-N$, $p^{\mu}(t,x,y)$ is continuous on $(0,\infty)\times K(|\mu|)$. By the symmetry propety, $p^{\mu}(t,x,y)$ is also continuous in (t,y) on $(0,\infty)\times K(|\mu|)$. We finally obtain (7.18) by the dominated convergence theorem and (7.15), completing the proof.

8. Estimates on the integral kernels

In this section we will restrict us to the case where $X = R^d$ and m is the Lebesgue measure. Let $\mu \in S - S$ such that $\mu^- \in C(\mu^+)$. Let $p^\mu(t,x,y)$ be as specified by Theorem 7.5.

<u>**8.1 Theorem**</u> Suppose that Assumption 2.5 holds. Then there exists positive constants c, β, λ such that

$$p^\mu(t,x,y) \le ce^{\beta t}\bar{p}(\lambda t,x,y), \forall t > 0, x,y \in X - N \tag{8.1}$$

where $\bar{p}(t,x,y) = (2\pi t)^{-\frac{d}{2}} e^{-\frac{|x-y|^2}{2t}}$ is the kernel associated to the Brownian motion and N is the exceptional set specified by (7.10).

<u>*Proof*</u> In what follows the C_i will always denote some positive constants. By Theorem 6.1 (iv) we have for m.a.e x

$$p^\mu(t,x,y) \le C_1 t^{-\frac{d}{2}}, \quad m.a.e.y, \forall 0 < t \le 1 \tag{8.2}$$

Applying Theorem 7.5 (i) and (ii) we see that (8.2) holds in fact for all $x \in X - N$ (with some new constant C_1. Fix such an $x \in X - N$ and denote by $B_\varepsilon(y)$ the d-dimensional cube centered at y and with volume $B_\varepsilon(y) = \varepsilon^d$. Let $\lambda > 1$ be such that $\lambda'\mu^- \in (\lambda'\mu^+)$ with $\frac{1}{\lambda'} + \frac{1}{\lambda} = 1$. We have

$$\frac{1}{|B_\varepsilon(y)|}\int_{B_\varepsilon(y)} p^\mu(t,x,z)m(dz) \le$$
$$\left[\frac{1}{|B_\varepsilon(y)|}E_x(e^{-\lambda'} A_t^\mu I_{B_\varepsilon(y)}(X_t))\right]^{\frac{1}{\lambda'}} \left[\frac{1}{|B_\varepsilon(y)|}E_x(I_{B_\varepsilon(y)}(X_t))\right]^{\frac{1}{\lambda}}$$

Applying (8.2) (w.r.t $p^{\lambda'\mu}(t,x,y)$) and letting $\mathcal{E} \downarrow 0$, we obtain for $t \le 1$ and $x \in X - N$, that

$$p^\mu(t,x,y) \le C_2\bar{p}(\lambda t,x,y), \text{m.a.e.}y$$

Therefore by Theorem 7.5 (ii) and (iii),

$$p^\mu(t,x,y) \le \int p^\mu(\frac{t}{2},x,z)p^\mu(\frac{t}{2},y,z)m(dz) \le C_3\bar{p}(\lambda t,x,y)$$
$$\forall 0 < t \le 1, \ x,y \in X - N \tag{8.3}$$

Put $c = \max\{C_3, 1\}, \beta = \log c$. Repeated application of (8.3) and Th. 7.5 (ii) implies the claimed assertion (8.1). ■

Acknowledgements

Two of the authors (S. Albeverio and Zhiming Ma) would like to thank Professors Hornung and Kotelenez for kind invitations to a very stimulating workshop in Oberwolfach. Very interesting discussions with Professor Masatoshi Fukushima and Michael Röckner are also gratefully acknowledged. We thank Dr. Walter Schneider for reading the manuscript and suggesting improvements and Carsten Welge, Heike Niehrenheim, Regina Kirchhoff and Bożena Zegarliński for their skilful help in the setting of the manuscript.

References

[ABRö] S. Albeverio, J. Brasche, M. Röckner: *Dirichlet forms and generalized Schrödinger operators*, in "Schrödinger operators", Proc. Sønderborg, Edts. H. Holden, A. Jensen, Lect. Note Phys. 345, Springer, Berlin (1989).

[AFHKL] S. Albeverio, J.E. Fenstad, R. Høegh-Krohn, T. Lindstrøm: *Nonstandard methods in stochastic analysis and mathematical physics*, Academic Press, Orlando (1986).

[AGKS] S. Albeverio, F. Gesztesy, W. Karwowski, L. Streit: *On the connection between Schrödinger- and Dirichlet forms*, J. Math. Phys. 26, 2546-2553 (1985).

[AHS] S. Albeverio, R. Høegh-Krohn, L. Streit: *Energy forms, Hamiltonians and distorted Brownian paths*, J. Math. Phys. 18, 907-917 (1977).

[AHK] S. Albeverio, R. Høegh-Krohn: *Quasi invariant measures, symmetric diffusion processes and quantum fields*, pp. 11–59 in "Les méthodes mathématiques de la théorie quantique des champs", Coll. Int. CNRS 248 (1976).

[AKS] S. Albeverio, S. Kusuoka, L. Streit: *Convergence of Dirichlet forms and associated Schrödinger operators*, J. Funct. Anal. 68, 130–148 (1986).

[AM1] S. Albeverio, Z. M. Ma:*Necessary and sufficient conditions for the existence of m - perfect processes associated with Dirichlet forms*, BiBoS-Preprint (1990).

[AM2] S. Albeverio, Z. M. Ma: *Nowhere Radon smooth measures, perturbations of Dirichlet forms and singular quadratic forms*, pp. 3-45 in Proc Bad Honnef Conf. 1988, ed. N. Christopeit, K. Helmes, M. Kohlmann Lect. Notes Control and Inform. Sciences 126, Springer, Berlin (1989).

[AM3] S. Albeverio, Z. M. Ma: *Additive functionals, nowhere Radon and Kato class smooth measure associated with Dirichlet forms*, SFB 237 - Preprint 66 (1989).

[AM4] S. Albeverio, Z. M. Ma: *Perturbation of Dirichlet forms - lower semiboundedness, closability and form cores*, SFB - 237 Preprint, to appear in J. Funct. Anal. (1990).

[AM5] S. Albeverio, Z. M. Ma: *Diffusion processes associated with singular Dirichlet forms*, to appear in Proc. Lisboa Conf., Ed. A. B. Cruzeiro (1990).

[ARö1] S. Albeverio, M. Röckner: *New developments in theory and applications of Dirichlet forms*, to appear in "Stochastic Processes, Physics and Geometry", Proc. 2nd Int. Conf. Ascona-Como- Locarno 1988, Edts. S. Albeverio, G. Casati, U. Cattaneo, D. Merlini, R. Moresi, World Scient., Singapore (1990).

[ARö2] S. Albeverio, M. Röckner: *Classical Dirichlet forms on topological vector spaces - closability and a Cameron-Martin formula*, J. Funct. Anal., 88, 395-436 (1990).

[ASi] M. Aizenman, B. Simon: *Brownian motion and Harnack's inequality for Schrödinger operators*, Commun. Pure Appl. Math. 35, 209-273 (1982).

[BCZ] Ph. Blanchard, Ph. Combe, W. Zheng: *Mathematical and Physical Aspects of Stochastic Mechanics*, Lect. Notes Phys. 281, Springer, Berlin (1987).

[BHH] A. Boukricha, H. Hueber, W. Hansen: *Continuous solutions of the generalized Schrödinger equation and perturbation of harmonic spaces* , Exp. Math. 5 97-135 (1987).

[BM1] Ph. Blanchard, Z. M. Ma: *Semigroup of Schrödinger operators with potentials given by Radon measures*, to appear in "Stochastic Processes - Physics and Geometry, Edt. S. Albeverio, G. Casati, U. Cattaneo, D. Merlini, R. Moresi, World Scient., Singapore (1990).

[BM2] Ph. Blanchard, Z. M. Ma: *Smooth measures and Schrödinger semigroups*, BiBoS 295 (1987).

[BM3] Ph. Blanchard, Ma, Z. M.: *New results on Schrödinger semigroups with potentials given by signed smooth measures*, To appear in Proc. Silivri Workshop 1988, Ed. Korezlioglu et al., Lect. Notes Maths., Springer (1989).

[Br] Z. Brzeźniak, *Some remarks on the Feynman - Kac formula* , J. Math. Phys. 31, 105-207 (1990).

[BrT] J.F. Brasche, A. Teta: *Spectral analysis and scattering theory for Schrödinger operators with an interaction supported by a regular curve*, SFB-Preprint No.73.

[C] J. A. van Casteren: *Generators of strongly continuous semigroups*, Research Notes in Math. 115, Pitman, London (1985).

[CKS] E. Carlen, S. Kusuoka, D.W. Stroock: *Upper bounds for symmetric Markov transition functions*, Ann. IHP 23 245-287 (1987).

[Da] E.B. Davies: *Heat kernels and spectral theory*, Cambridge University Press, 1989.

[DC] M. Demuth, J. A. von Casteren: *On spectral theory of selfadjoint Feller generators*, Uni. Antwerpen Preprint (1989), to appear in Rev. Math. Phys. (1990).

[Ch] K.L. Chung, *Probabilistic approach to boundary value problems for Schrödinger's equation* Exp. Math. 3, 175-178 (1985).

[Che] S.E. Cheremshantsev: *Hamiltonians with zero range interactions supported by Brownian paths*, Leningrad Preprint (1989).

[DGS] B. Davies, L. Gross, B. Simon: *Hypercontractivity: a bibliographic review*, to appear in Mem. Volume R. Høegh-Krohn, Cambridge University Press (1991).

[DS] E.B. Davies, B. Simon: *Ultracontractivity and the heat kernel for Schrödinger operators and Dirichlet Laplacians*, J. Funct. Anal. 59, 335-395 (1985).

[F1] M. Fukushima: *Dirichlet forms and Markov processes*, Kodanska and North Holland, (1980).

[F2] M. Fukushima: *Energy forms and diffusion processes*, pp. 65-98 in "Mathematics and Physics", Vol. 1. Ed. L. Streit, World Scient., Singapore (1985).

[F3] M. Fukushima: *On a stochastic calculus related to Dirichlet forms and distorted Brownian motion*, Phys. Rep. 77 (1981).

[F4] M. Fukushima: *On two classes of smooth measures for symmetric Markov processes*, Osaka University Preprint.

[FO] M. Fukushima, Y. Oshima: *On skew product of symmetric diffusion processes*, Forum Math. 1, 103-142, (1989).

[Fr] A. Friedman: *Stochastic differential equations and applications*, Vol. 1, Academic Press, New York, London (1976).

[J] G. W. Johnson: *Existence theorems for the analytic operator-valued Feynman integral*, Séminaire d'analyse moderne, No. 20, Université de Sherbrooke, 1988.

[Ma] Z.M. Ma: *Some new results concerning Dirichlet forms, Feynman–Kac semigroups and Schrödinger equations*, to appear in Contemp. Maths., AMS (1990).

[MaS] Z. M. Ma, R.M. Song, *Probabilistic approach to Schrödinger equation*, Sem. Stoch. Processes, Ed. K.L. Chung et al., Birkhäuser (1989).

[Mc] Mc Kean: *The Bessel motion and a singular integral equation*, Mem. College of Sciences, University of Kyoto, Theory A, 33, Math. No 2, 317-322 (1960).

[Pa1] D. Pantić, *Stochastic calculus on distorted Brownian motion* , J. Math. Phys. 29, 207 -209 (1988).

[Pa2] D. Pantić, *Stochastic calculus and Schrödinger semigroups admitting a ground state*, J. Math. Phys. 31, 1221-1225 (1990)

[RS1] M. Reed, B. Simon: *Methods of Modern Mathematical Physics (I)*, Revised and enlarged edition, Academic Press, New York (1980).

[RS2] M. Reed, B. Simon: *Methods of Modern Mathematical Physics (II)*, Revised and enlarged edition, Academic Press, New York (1980).

[S1] M. L. Silverstein: *Symmetric Markov Processes*, Lect. Notes Maths. 426, Springer, Berlin (1974).

[S2] M. L. Silverstein: *Boundary Theory for Symmetric Markov Processes*, Lect. Notes Maths. 516, Springer Berlin (1976).

[Sch] H. H. Schaefer: *Topological vector spaces*, MacMillan (1987).

[Sim1] B. Simon: *Schrödinger semigroups*, Bull. Am. Math. Soc. 7, 447-526 (1982).

[Sim2] B. Simon: *Functional Integration and Quantum Physics*, Academic Press, New York, London (1979).

[St] T. Sturm: *Störung von Hunt Prozessen durch signierte additive Funktionale*, Ph. D Thesis, Erlangen-Nürnberg, (1989).

[V] J. Voigt: *Absorption Semigroups. Their Generators and Schrödinger semigroups*, Functional Analysis, Vol. 67, 165-205 (1986).

[Va] N. Th. Varopoulos: *Hardy-Littlewood theory for semigroups*, J. Funct. Anal. 63, 240-260 (1985).

[Zh] Z.X. Zhao: *Conditional gauge with unbounded potentials*, Zeitschr. Wahrsch. verw. Geb. 65 (1983), 13-18.

FLUCTUATION FIELD FOR THE ASYMMETRIC SIMPLE EXCLUSION PROCESS (1)

Albert Benassi
CNRS. Laboratoire de Probabilités, Université de Paris 6
4, place Jussieu, Tour 56, F-75252 PARIS CEDEX 05

Jean-Pierre Fouque
CNRS. Centre de Mathématiques Appliquées
Ecole Polytechnique, F-91128 PALAISEAU CEDEX

Abstract
The hydrodynamical behavior of one-dimensional asymmetric attractive particle systems is described by a first order non linear hyperbolic partial differential equation. This corresponds to a law of large numbers for the system and it is natural to study the fluctuations. When the density of particles is zero before the shock, we show that the fluctuation field is a solution, in a weak sense, of a linear equation with discontinuous coefficients, namely the linearization of the hydrodynamical equation. The fluctuation of the shock itself, studied in [8], will be our main tool.

INTRODUCTION AND NOTATIONS

We will only discuss the case of the one-dimensional nearest neighbours completely asymmetric exclusion process which can be described as follows : at each site k of **Z** there is at most one particle, each particle waits an exponential time with parameter one and jumps at site k+1 if it is empty and remains at site k if it is occupied. This is a markovian evolution on the space of configurations $\eta \in \{0,1\}^{\mathbf{Z}} = E$ with the infinitesimal generator L acting on cylindrical functions by :

$$Lf(\eta) = \sum_{k \in \mathbf{Z}} \eta(k)\left(1-\eta(k+1)\right)\left[f(\eta^{k,k+1}) - f(\eta)\right]$$

where $\eta^{k,k+1}$ is obtained from η by exchanging sites k and k+1. We will denote by $(T)_{t\geq 0}$ the associated semigroup and by τ_k the shift which acts on E by $\tau_k\, \eta(\ell) = \eta(\ell+k)$ or on functions by $(\tau_k f)(\eta) = f(\tau_k \eta)$ or on probability measures on E by $\int fd(\tau_k\mu) = \int (\tau_k f)\, d\mu$.

(1) Presented by the second author at the Oberwolfach Conference.

We will denote by ν^ρ the equilibrium measure at density $0 \leq \rho \leq 1$, product of Bernoulli measures such that $\nu^\rho\{\eta/\eta(k)=1\} = \rho$ for every $k \in \mathbf{Z}$. Our initial distribution will be a product measure μ_ρ such that $\mu_\rho\{\eta/\eta(k)=1\} = 0$ if $k<0$ and ρ if $k \geq 0$.

We will denote by x_t the position, at time t, of the first particle : $x_t = \inf\{k \in \mathbf{Z}/\eta_t(k)=1\}$. We know by the basic coupling, that we can add ξ-particles on $\mathbf{Z}$- in such a way that η-particles are initially distributed according to μ_ρ and evolve according to L and $(\eta+\xi)$-particles are initially distributed according to ν^ρ and evolve according to L. Since particles are jumping only to the next neighbour (to the right), the first η-particle is an interface between η and ξ-particles.

In [1], we proved *local equilibrium* :

$$\tau_{[x/\varepsilon]}\, \mu_\rho T_{t/\varepsilon} \underset{\varepsilon \downarrow 0}{\overset{\text{weakly}}{\longrightarrow}} \nu^{u(x,t)}$$

where $x \in \mathbf{R} \backslash \{(1-\rho)t\}$, $[x/\varepsilon]$ is the integer part of x/ε and $u(x,t)$ is the *unique entropic solution* of the *hydrodynamical equation* :

$$(*) \qquad \begin{cases} \dfrac{\partial u}{\partial t} + \dfrac{\partial(u(1-u))}{\partial x} = 0 \\ u(x,0) = u_0(x) = \rho\, \mathbb{1}_{[0,+\infty)}(x) \end{cases}$$

In fact we have propagation of the shock, $u(x,t) = \rho\, \mathbb{1}_{[(1-\rho)t,+\infty)}(x)$, and more precisely we were able to prove that :

$$\varepsilon \sum_{k=[x/\varepsilon]}^{[y/\varepsilon]} \eta_{t/\varepsilon}(k) \underset{\varepsilon \downarrow 0}{\overset{\text{a.s.}}{\longrightarrow}} \int_x^y u(z,t)\, dz$$

or in a functional form : for every continuous function φ with compact support :

$$\varepsilon \sum_{k \in \mathbf{Z}} \varphi(\varepsilon k)\, \eta_{t/\varepsilon}(k) \underset{\varepsilon \downarrow 0}{\overset{\text{a.s.}}{\longrightarrow}} \int_{\mathbf{R}} \varphi(x)\, u(x,t)\, dx.$$

A presentation of this result with simulations can be found in [5].

The convergence to (*) has been studied in various situations for attractive particle systems.

In [2] we extended this result to general initial distributions u_0.

At the shock, it has been proved that :

(1) $$\tau_{[(1-\rho)t/\varepsilon]}\, \mu_\rho T_{t/\varepsilon} \underset{\varepsilon\downarrow 0}{\overset{\text{weakly}}{\longrightarrow}} \frac{1}{2}(\nu^0+\nu^\rho)$$

(2) $$\varepsilon\ x_{t/\varepsilon} \underset{\varepsilon\downarrow 0}{\overset{\text{a.s.and } L^1}{\longrightarrow}} (1-\rho)t$$

(3) $$\frac{1}{\sqrt{\varepsilon}}\left(\varepsilon\ x_{t/\varepsilon} - (1-\rho)t\right) \underset{\varepsilon\downarrow 0}{\overset{\text{in law}}{\longrightarrow}} \mathcal{N}(0,t\sigma^2>0)$$

(see [8], [4] and [7]).

Our purpose is to define the fluctuation field

$$Y_t^\varepsilon(x) = \frac{1}{\sqrt{\varepsilon}}\left\{\eta_{t/\varepsilon}([x/\varepsilon])-u(x,t)\right\}$$

and to study its asymptotic behavior as $\varepsilon\downarrow 0$.

At $t=0$, the limiting field Y^0 is gaussian, centered with covariance defined by :

$$\lim_{\varepsilon\downarrow 0} \int Y_0^\varepsilon(\varphi)\, Y_0^\varepsilon(\psi)d\mu_\rho = \int_{\mathbf{R}} \varphi(x)\psi(x)u_0(x)\left(1-u_0(x)\right) dx$$

$$= \int_0^\infty \varphi(x)\psi(x)\rho(1-\rho)\, dx$$

where, for a smooth function φ as $\varphi\in \mathcal{S}(\mathbf{R})$,

$$Y_t^\varepsilon(\varphi) = \int_{\mathbf{R}} Y_t^\varepsilon(x)\ \varphi(x)\ dx.$$

We will show that the limiting field Y_t is a solution in a weak sense, away from the shock, of the following equation :

$$(**)\quad \begin{cases} \dfrac{\partial Y}{\partial t} + \dfrac{\partial\big(((1-2u(x,t))Y\big)}{\partial x} = 0 \\ \\ Y_0 = Y^0 \end{cases}$$

which is the linearization of (*).

In fact we are studying the *conservative system*

$$(S_c)\quad \begin{cases} \dfrac{\partial u}{\partial t} + \dfrac{\partial\big(u(1-u)\big)}{\partial x} = 0 \\ \\ \dfrac{\partial Y}{\partial t} + \dfrac{\partial\big((1-2u)Y\big)}{\partial x} = 0 \end{cases}$$

and the difficulty comes from the fact that, (1-2u) being discontinuous and Y not a function, the product (1-2u)Y does not make sense in general (see [9] for a treatment of this system in the case Y is measure-valued). By integrating Y^ε, namely $Z_t^\varepsilon = \int_{-\infty}^{x} Y_t^\varepsilon(y)\, dy$, we obtain the following *non conservative system :*

$$(S_{nc})\quad \begin{cases} \dfrac{\partial u}{\partial t} + \dfrac{\partial\big(u(1-u)\big)}{\partial x} = 0 \\ \\ \dfrac{\partial Z}{\partial t} + (1-2u)\dfrac{\partial Z}{\partial x} = 0 \end{cases}$$

which will be studied in a coming article in the case $Z_0(x) = 0$ for $x<0$ and $(Z_0(x))_{x\geq 0}$ is a brownian motion starting from 0.

Its solution can be written :

$$Z_t(x) = \mathbb{1}_{[(1-\rho)t,+\infty]}(x)\, Z_0(x-(1-2\rho)t).$$

In this article we show that the fluctuation field, *away from the shock*, is gaussian and is, in fact, the initial noise transported by (**) *without additional noise* coming from the evolution in time. We would like to mention the fact that, in the case of the symmetric exclusion process, the situation is radically different : the appropriate scaling in time is

t/ε^2, the hydrodynamical equation is parabolic, namely the heat equation, and the fluctuation field is solution of a parabolic stochastic P.D.E. with additional noise coming from the evolution (cf [3]).

RESULTS.

THEOREM 1 - For fixed $t>0$, Y_t^ε *converges in law to a centered gaussian field* Y_t *on* $\mathbf{R}\backslash\{(1-\rho)t\}$ *with covariance given by :*

$$E\{Y_t(\varphi)\, Y_t(\psi)\} = \int_{(1-\rho)t}^{+\infty} \varphi(x)\, \psi(x)\rho(1-\rho)\, dx$$

where φ *and* ψ *are* C^∞ *functions with compact supports in* $\mathbf{R}\backslash\{(1-\rho)t\}$.

Proof. Denote by $X_t^\varepsilon(x)$ the fluctuation field associated with $(\xi + \eta)$-particles : $X_t^\varepsilon(x) = \frac{1}{\sqrt{\varepsilon}}\{\eta_{t/\varepsilon}([x/\varepsilon]) + \xi_{t/\varepsilon}([x/\varepsilon]) - \rho\}$. This field (under the equilibrium ν^ρ) converges in law to a centered gaussian field X with covariance :

$$E\{X(\varphi)X(\psi)\} = \int_R \varphi(x)\, \psi(x)\, \rho(1-\rho)\, dx$$

where φ and ψ belong to $\mathcal{D}(\mathbf{R})$ or $\mathcal{S}(\mathbf{R})$.

For φ in $\mathcal{D}(\mathbf{R}\backslash\{(1-\rho)t\})$ let $\delta>0$ such that $\operatorname{supp}\varphi \cap [(1-\rho)t - \delta, (1-\rho)t + \delta] = \phi$ and denote by $A_{t,\delta}^\varepsilon$ the set $\{|\varepsilon x_{t/\varepsilon} - (1-\rho)t| < \delta\}$.

From the definition of $x_{t/\varepsilon}$ and its interface property we get that :

$$Y_t^\varepsilon(\varphi) = \begin{cases} 0 & \text{if } \operatorname{supp}\varphi \subset (-\infty,(1-\rho)t) \\ X_t^\varepsilon(\varphi) & \text{if } \operatorname{supp}\varphi \subset ((1-\rho)t,+\infty) \end{cases}$$

on the set $A_{t,\delta}^\varepsilon$. From (2) we have that $\lim_{\varepsilon\downarrow 0} P(A_{t,\delta}^\varepsilon) = 1$ and we deduce the result. □

Remark : In the case of this completely asymmetric dynamics and with this initial distribution μ_ρ (no particle on $\mathbf{Z}$-), this argument gives a direct proof of the local equilibrium property :

$$\{\eta_{t/\varepsilon}([x/\varepsilon]+k),\, k\in \mathbf{Z}\} \underset{\varepsilon\downarrow 0}{\overset{\text{weakly}}{\longrightarrow}} \nu^{u(x,t)}$$

with $u(x,t) = \rho\, \mathbb{1}_{[(1-\rho)t,+\infty)}(x)$, for $x \neq (1-\rho)t$.

<u>*THEOREM 2*</u> - *The same convergence holds for*

$$\widetilde{Y}_t^\varepsilon(x) = \frac{1}{\sqrt{\varepsilon}}\{\eta_{t/\varepsilon}([x/\varepsilon]) - E(\eta_{t/\varepsilon}([x/\varepsilon]))\}.$$

<u>Proof</u> - We have :

$$\widetilde{Y}_t^\varepsilon(x) - Y_t^\varepsilon(x) = \frac{1}{\sqrt{\varepsilon}}\{u(x,t) - E(\eta_{t/\varepsilon}([x/\varepsilon]))\}$$

$$= \begin{cases} \frac{1}{\sqrt{\varepsilon}} E(\xi_{t/\varepsilon}([x/\varepsilon])) \text{ if } x>(1-\rho)t \\ -\frac{1}{\sqrt{\varepsilon}} E(\eta_{t/\varepsilon}([x/\varepsilon])) \text{ if } x<(1-\rho)t \end{cases}$$

If $\varphi\in \mathcal{D}(\mathbf{R}\backslash\{(1-\rho)t\})$ and $\operatorname{supp}\varphi \cap [(1-\rho)t-\delta, (1-\rho)t+\delta] = \phi$ for some $\delta>0$, we have :

$$\left|\widetilde{Y}_t^\varepsilon(\varphi) - Y_t^\varepsilon(\varphi)\right| \leq \frac{1}{\sqrt{\varepsilon}} P\{\left|\varepsilon x_{t/\varepsilon} - (1-\rho)t\right| > \delta\}.$$

By (2), for ε small enough, $\left|E(\varepsilon x_{t/\varepsilon}) - (1-\rho)t\right| < \frac{\delta}{2}$ and then $P\{|\varepsilon x_{t/\varepsilon} - (1-\rho)t| > \delta\} \leq P\{\left|\varepsilon x_{t/\varepsilon} - E(\varepsilon x_{t/\varepsilon})\right| > \frac{\delta}{2}\} \leq \frac{\operatorname{Var}(\varepsilon x_{t/\varepsilon})}{\delta^2/4}$ which is equivalent, by (3), to $\frac{4t\,\sigma^2\,\varepsilon}{\delta^2}$; therefore $\left|\widetilde{Y}_t^\varepsilon(\varphi) - Y_t^\varepsilon(\varphi)\right|$ converges to 0 and we deduce the result. □

The next result justifies the equation (**) for the limiting field Y_t.

We can write a stochastic equation for $Y_t^\varepsilon(x)$; we denote by $\{P_t(k),\, k\in \mathbf{Z}\}$ a family of independent Poisson processes with intensity 1 ; we have :

$$\begin{aligned} d\eta_t(k) \;=\; & -\eta_{t^-}(k)(1-\eta_{t^-}(k+1))\, dP_t(k) \\ & +\eta_{t^-}(k-1)(1-\eta_{t^-}(k))\, dP_t(k-1) \end{aligned}$$

and therefore we get an equation for $\eta_{t/\varepsilon}([x/\varepsilon])$ by replacing k by $[x/\varepsilon]$ and t by t/ε.

Using the notations : $D_{-\varepsilon}u(x) = -\frac{1}{\varepsilon}(u(x-\varepsilon) - u(x))$ and $F_\varepsilon u(x) = u(x)\ (1 - u(x+\varepsilon))$ we get :

$$d\eta_{t/\varepsilon}([x/\varepsilon]) = - D_{-\varepsilon}\left\{F_\varepsilon\left(\eta_{t/\varepsilon^-}([x/\varepsilon])\right) d\left(\varepsilon P_{t/\varepsilon}([x/\varepsilon])\right)\right\}$$

which can be written in the weak form :

$$\int_{\mathbf{R}}\int_{\mathbf{R}_+} \left\{\eta_{t/\varepsilon}([x/\varepsilon])\frac{\partial\varphi}{\partial t}\, dt + (D_\varepsilon\varphi)\, F_\varepsilon\left(\eta_{t/\varepsilon^-}([x/\varepsilon])\right) d\left(\varepsilon P_{t/\varepsilon}([x/\varepsilon])\right)\right\} dx = - \int_{\mathbf{R}} \eta_0([x/\varepsilon])\, \varphi(x,0)\, dx \tag{4}$$

for any smooth function φ with compact support in $\mathbf{R}\times\mathbf{R}_+$. Taking the expectation, we get :

$$\int_{\mathbf{R}}\int_{\mathbf{R}_+} \left\{E\left(\eta_{t/\varepsilon}([x/\varepsilon])\right)\frac{\partial\varphi}{\partial t} + (D_\varepsilon\varphi)\, E\left(F_\varepsilon\left(\eta_{t/\varepsilon}([x/\varepsilon])\right)\right)\right\} dt\, dx = - \int_{\mathbf{R}} u_0(x)\varphi(x,0)\, dx \tag{5}$$

Substracting (5) to (4) and dividing by $\sqrt{\varepsilon}$ we get an equation for $\widetilde{Y}_t^\varepsilon$:

$$\int_{\mathbf{R}}\int_{\mathbf{R}_+} \left\{\widetilde{Y}_t^\varepsilon\frac{\partial\varphi}{\partial t}dt + (D_\varepsilon\varphi)\frac{F_\varepsilon\left(\eta_{t/\varepsilon^-}([x/\varepsilon])\right)d\left(\varepsilon P_{t/\varepsilon}([x/\varepsilon])\right) - E\left(F_\varepsilon\left(\eta_{t/\varepsilon}([x/\varepsilon])\right)\right)dt}{\sqrt{\varepsilon}}\right\}dx = - \int_{\mathbf{R}} Y_0^\varepsilon(x)\varphi(x,0)\, dx \tag{6}$$

Writing (*) in the weak form and substracting from (4) we also get an equation for Y_t^ε :

$$\int_{\mathbf{R}}\int_{\mathbf{R}_+} \left\{Y_t^\varepsilon\frac{\partial\varphi}{\partial t}dt + (D_\varepsilon\varphi)\frac{F_\varepsilon\left(\eta_{t/\varepsilon}([x/\varepsilon])\right)d\left(\varepsilon P_{t/\varepsilon}([x/\varepsilon])\right) - u(x,t)\left(1-u(x,t)\right)dt}{\sqrt{\varepsilon}}\right\}dx = - \int_{\mathbf{R}} Y_0^\varepsilon(x)\varphi(x,0)\, dx \tag{7}$$

LEMMA 1

$$E\left|\int_{R}\int_{R_+}(D_\varepsilon\varphi)F_\varepsilon(\eta_{t/\varepsilon}([x/\varepsilon]))\left\{\frac{d(\varepsilon P_{t/\varepsilon}([x/\varepsilon]))-dt}{\sqrt{\varepsilon}}\right\}dx\right|^2 \underset{\varepsilon\downarrow 0}{\to} 0.$$

In other words, there will be no additional noise in the limiting equation due to the evolution in time.

Proof - The double integral we have to compute is equal to :

$$\int_{R_+}\sum_{k\in Z}\left(\int_{\varepsilon k}^{\varepsilon(k+1)}(D_\varepsilon\varphi)\,F_\varepsilon(\eta_{t/\varepsilon^\cdot}(k))dx\right)\frac{d(\varepsilon P_{t/\varepsilon}(k))-dt}{\sqrt{\varepsilon}}\,.$$

$\varepsilon P_{t/\varepsilon}(k) - t = \varepsilon\, M_{t/\varepsilon}(k)$ is a family of orthogonal martingales with increasing processes εt ; since
$|F_\varepsilon(\eta(k))| \leq 1$,

$$\left|\int_{\varepsilon k}^{\varepsilon(k+1)}(D_\varepsilon\varphi)\,F_\varepsilon(\eta_{t/\varepsilon^\cdot}(k)\,dx\right| \leq \varepsilon\,||\varphi'||_\infty.$$

If T>0 and A>0 are such that supp $\varphi \subset [-A,A]\times[0,T]$, the quantity we have to estimate is less than :

$$\int_0^T \sum_{\substack{k/\\ |\varepsilon k|\leq A+1}} \varepsilon^2||\varphi'||_\infty^2\frac{\varepsilon dt}{(\sqrt{\varepsilon})^2} \sim T\sum_{k=-[(A+1)/\varepsilon]}^{[(A+1)/\varepsilon]}\varepsilon^2\,||\varphi'||_\infty^2 \sim \text{cste}\times\varepsilon$$

which proves the lemma. □

If the initial distribution is ν^ρ, (6) or (7) give the same equation for X_t^ε which, obviously, converges weakly as a continuous $\mathcal{D}$'(R)–valued process to (X_t) which will be solution to :

$$(**)\begin{cases}\dfrac{\partial X}{\partial t}+\dfrac{\partial((1\text{-}2\rho)X)}{\partial x}=0\\[2ex] X_0\end{cases}$$

where X_0 is the centered gaussian field with covariance
$E\{X_0(\varphi)\, X_0(\psi)\} = \int_{\mathbf{R}} \varphi(x)\, \psi(x)\, \rho(1-\rho)dx.$

In that case we have the unicity of the solution since the coefficient is constant ; one can write :
$X_t = \tau_{-(1-2\rho)t}\, X_0$ where τ is the space shift. The coefficient $1-2\rho = \frac{d}{d\rho}\,(\rho(1-\rho))$ comes from the simple computation :

$$\frac{F_\varepsilon(\eta_{t/\varepsilon}([x/\varepsilon]))-\rho(1-\rho)}{\sqrt{\varepsilon}} = X_t^\varepsilon(x)-\rho X_t^\varepsilon(x)-\rho\tau_\varepsilon X_t^\varepsilon(x)$$

$$- X_t^\varepsilon(x)\big(\eta_{t/\varepsilon}([x/\varepsilon]+1)-\rho\big)$$

Doing the same computation for Y_t^ε, we get :
1) if supp $\varphi \subset \{(x,t) \,/\, x > (1-\rho)t\}$, $u(x,t) = \rho$ and

$$\frac{F_\varepsilon(\eta_{t/\varepsilon}([x/\varepsilon]))-\rho(1-\rho)}{\sqrt{\varepsilon}} = Y_t^\varepsilon(x)-\rho Y_t^\varepsilon(x)-\rho\tau_\varepsilon Y_t^\varepsilon(x)$$

$$- Y_t^\varepsilon(x)\big(\eta_{t/\varepsilon}([x/\varepsilon]+1)-\rho\big) - \frac{E\{\eta_{t/\varepsilon}([x/\varepsilon])\eta_{t/\varepsilon}([x/\varepsilon]+1)\} - \rho^2}{\sqrt{\varepsilon}}$$

2) if supp $\varphi \subset \{(x,t) \,/\, x < (1-\rho)t\}$, $u(x,t) = 0$ and

$$\frac{F_\varepsilon(\eta_{t/\varepsilon}([x/\varepsilon]))}{\sqrt{\varepsilon}} = Y_t^\varepsilon(x)-Y_t^\varepsilon(x)\, \eta_{t/\varepsilon}([x/\varepsilon])$$

$$- \frac{E\{\eta_{t/\varepsilon}([x/\varepsilon])\eta_{t/\varepsilon}([x/\varepsilon]+1)\}}{\sqrt{\varepsilon}}$$

The techniques of the interface $x_{t/\varepsilon}$ used in theorem 2 improved as follows :
$\left\{\frac{1}{\sqrt{\varepsilon}}\big(\varepsilon\, x_{t/\varepsilon} - (1-\rho)t\big)\right\}_{t\geq 0}$ converges weakly to a real brownian motion $(B_t)_{t\geq 0}$ with $B_0=0$ and $E(B_t^2) = \sigma^2 t$; and the fact that $\int Y_0^\varepsilon(x)\, \varphi(x,0)\, dx$ converges in law to a centered gaussian variable with covariance $\int_{\mathbf{R}} (\varphi(x))^2\, \rho(1-\rho)\, dx$ in the first case and to 0 in the second case, enable us to give the final result :

THEOREM 3 - $(Y_t^\varepsilon(x))$ *converges in law to a* $\mathcal{D}'(\{x \neq (1-\rho)t\}$*-valued field* Y *which is a weak solution on* $\{x\neq(1-\rho)t\}$ *of the equation* (**) *starting from the initial centered* *gaussian field* Y_0.

Remark - Using equation (6) in place of equation (7), with the same method we get the same convergence for $(\widetilde{Y}_t^{\varepsilon}(x))$.

To summarize, we have $Y_t = 0$ on $\{x < (1-\rho)t\}$ and $Y_t = \tau_{-(1-2\rho)t} Y_0$ on $\{x > (1-\rho)t\}$.

COMMENTS

1) For the convergence of $(X_t^{\varepsilon})_{t\geq 0}$ to $(X_t)_{t\geq 0}$ as a continuous process taking its values in $\mathcal{D}'(\mathbf{R})$, we use (6) to get the tightness, and the unicity of the solution of the limiting equation. Obviously we could prove this convergence on a smaller space than $\mathcal{D}'(\mathbf{R})$, namely the Sobolev space carrying X_0.

2) We did not say anything at the shock $x = (1-\rho)t$. We refer to [7] and [4] for a description of the microscopic fluctuations of the shock.

3) In the absence of the complete asymmetry hypothesis, we lose the microscopic interface property and our argument does not work directly. Still the result under the equilibrium ν^{ρ} is true and the use of [4] should give the result under μ_{ρ}.

4) In order to study the fluctuation field in the case of an initial product distribution $\mu_{\lambda,\rho}$ such that : $\mu_{\lambda,\rho}\{\eta/\eta(k) = 1\} = \lambda$ if $k < 0$ and ρ if $k \geq 0$, with $0<\lambda<\rho$, we need to know the fluctuations of the interface for a second class of particles which seems to be an unsolved difficult problem.

5) In the case of a rarefaction wave $(\lambda > \rho)$, the situation is not clear at all since between the characteristics $x = (1-\lambda)t$ and $x = (1-\rho)t$, the solution of (**) is not defined.

For instance, in the case $\lambda = 1$ and $\rho = 0$, there is no initial noise but one is certainly present at $t > 0$, between these characteristics !!

REFERENCES

[1] A. BENASSI, J.P. FOUQUE (1987).- *Hydrodynamical limit for the asymmetric simple exclusion process*.- Ann. Proba. 15, 546-560.

[2] A. BENASSI, J.P. FOUQUE, E. SAADA, M.E. VARES (1990).- *One dimensional asymmetric attractive particle systems : hydrodynamical limit for general initial conditions*.- (Preprint).

[3] A. DE MASI, N. IANIRO, A. PELLEGRINOTTI, E. PRESUTTI (1984). *A survey of the hydrodynamical behavior of many particle systems*.- North Holland (J.L. Lebowitz and E.W. Montroll, eds), 123-294.

[4] A. DE MASI, C. KIPNIS, E. PRESUTTI, E. SAADA (1989).- *Microscopic structure at the shock in the asymmetric simple exclusion*.- Stochastics, 27, 151-165.

[5] J.P. FOUQUE (1990).- *Hydrodynamical behavior of asymmetric particle systems. One example : one dimensional nearest neighbours asymmetric simple exclusion process*.- AMS-SIAM Summer Seminar (to appear).

[6] J.P. FOUQUE (1984).- *La convergence en loi pour les processus à valeurs dans un espace nucléaire*.- Ann. de l'I.H.P., 20, 3, 225-245.

[7] J. GARTNER, E. PRESUTTI (1989).- *Shock fluctuations in a particle system*.- (Preprint).

[8] C. KIPNIS (1985).- *Central limit theorems for infinite series of queues and applications to simple exclusion*.- Ann. of Proba., 13, 397-408.

[9] Ph. LE FLOCH (1989).- *An existence and uniqueness result for two non strictly hyperbolic systems*. (Preprint).

[10] D. WICK (1985).- *A dynamical phase transition in an infinite particle system*.- J. Stat. Phys., 38, 1015-1025.

ALMOST SURE CONVERGENCE OF SOME APPROXIMATE SOLUTIONS FOR RANDOM PARABOLIC EQUATIONS *

Pao-Liu Chow and Jing-Lin Jiang

Department of Mathematics, Wayne State University

Detroit, Michigan 48202, U.S.A.

ABSTRACT

The paper is concerned with some approximate solutions for random partial differential equations of parabolic type via discretization in time or by an eigenfunction expansion. In contrast with L^p-convergence, it is shown that, under suitable conditions, the approximate solutions converge almost surely to the strong solution of a parabolic Itô equation.

1. Introduction

Let us consider a random parabolic equation in domain $D \subset \mathbf{R}^d$ as follows:

$$\frac{\partial u(t,x)}{\partial t} = \Delta u + f(x,u) + \sigma(t,x)\dot{w}_t(x) , \tag{1.1}$$

which is subject to the initial and boundary conditions:

$$u(0,x) = v(x) , \tag{1.2}$$

$$u(t,x)|_{\partial D} = 0 . \tag{1.3}$$

Here Δ is the Laplacian operator; $f(x,u)$, $\sigma(t,x)$ and $v(x)$ are some given functions; ∂D denotes the smooth boundary of D, and $\dot{w}_t(x) = \frac{\partial}{\partial t}w_t(x)$ is formally a white noise at point x. The random field $w_t(x)$ is a Wiener process, that is, a centered Gaussian process with covariance function $r(x,y)$ defined by

$$Ew_t(x)w_s(y) = (t \wedge s)r(x,y), \quad x,y \in D, \tag{1.4}$$

where $t \wedge s = \min\,(t,s)$.

*The work of the first author was supported by NSF grant # DMS-87-02236.

In a Hilbert space setting, the system (1.1)–(1.3) can be written as an Itô equation of the form:

$$(1.5)\qquad \begin{cases} du_t = [Au_t + F(u_t)]dt + \Sigma_t dW_t \ , \\ u_0 = v \ , \end{cases}$$

where $u_t = u(t,\cdot)$ and $v = v(\cdot)$ are regarded as vectors in $H = L^2(D)$, $Av = \Delta v(\cdot)$, $F(v) = f(\cdot, v)$, $\Sigma_t = \sigma(t,\cdot)$ and $W_t = w_t(\cdot)$ is a H-valued Brownian motion. Rewriting (1.5) as an Itô integral equation

$$(1.6)\qquad u_t = v + \int_0^t B(u_s)ds + \int_0^t \Sigma_s dW_s \ ,$$

with $B(v) = Av + F(v)$, the existence of a strong solution was established by Pardoux [1] and others. As an ordinary Itô equation, one would naturally think of salving Equation (1.6) successively by Euler's method based on a time-discretization. In finite dimensions, this problem was treated, for instance, in Ikeda and Watanabe [2]. On the other hand, as PDE's (partial differential equations), the method of eigenfunction expansion or, more generally, the Galerkin method is often used to construct an approximate solution. In fact this provides a constructive way of proving the existence theorem, as done in [1]. As far as we know, in the case of random PDE's, all convergence results were proved in the sense of L^p or weak convergence. The question of almost sure or pathwise convergence has not yet been dealt with.

In this paper we will present two results on the a.s. convergence of approximate solutions to a random PDE of the form (1.5). They correspond, respectively, to the approximation schemes based on a time-discretization and the eigenfuction expansion. For simplicity, we have assumed that the noise is independent of state, i.e. Σ in (1.5) is not a function of u. The general problem, that includes the case of state-dependent noise, will be discussed in detail elsewhere.

2. Convergence of Discrete-time approximations

Let H be a real separable Hilbert space with inner product $(\cdot,\cdot)$ and norm $|\cdot|$, and let $\{W_t,\ t \geq 0\}$ be a H-valued Wiener process with mean zero and covariance operator R defined by

$$E(W_t, g)(W_s, h) = (t \wedge s)(Rg, h), \quad \forall\, g, h \in H.$$

For convenience, as in Chow [3], let us introduce a subspace $H_R \subset H$, the closure of $R^{1/2}H$ with respect to the norm $|\cdot|_R$ defined by $|R^{1/2}h|_R^2 = (R^{1/2}h, R^{1/2}h)_R = (h,h)$.

Let $V \subset H$ be a dense Sobolev subspace. Suppose that $A : V \to V'$, the dual of V, and $\Sigma_t : H_R \to H$ are linear operators, and $F : H \to H$ is bounded. We assume that

(A.1) $A : V \to V'$ is coercive such that it generates an analytic semigroup $\{T_t, t \geq 0\}$ on H.

(A.2) F satisfies the following boundedness and Lipschitz conditions:

$$|F(h)|^2 \leq L(1 + |h|^2) , \quad \forall h \in H$$

and

$$|F(g) - F(h)| \leq L|g - h| , \quad \forall g, h \in H$$

where $L > 0$ is a constant.

(A.3) $\Sigma : [0, \ell] \to \mathcal{L}(H_R, H)$ is bounded and Hölder-continuous with exponent ν such that

$$\|A^\alpha \Sigma_t\|_R \leq L$$

and

$$\|\Sigma_t - \Sigma_s\|_R \leq L|t - s|^\nu,$$

for some constants $L > 0$, α and $\nu > 0$, where $\|\cdot\|_R$ denotes the operator norm in $\mathcal{L}(H_R, H)$, the space of bounded linear operators from H_R into H.

Under the above conditions, the equation (1.5) is known to have a unique strong solution [1]. This solution is also a mild solution which satisfies the following integral equation:

$$u_t = T_t v + \int_0^t T_{t-s} F(u_s) ds + \int_0^t T_{t-s} \Sigma_s dW_s \,, \tag{2.1}$$

with $v \in H$.

Now we consider approximate solutions based on a time-discretization scheme. Partition the interval $[0, \ell]$ into n subintervals of equal length. Let us define

$$\pi_n s = \frac{i}{n}\ell \quad \text{if} \;\; s \in [\frac{i}{n}\ell, \frac{i+1}{n}\ell), \quad i = 0, 1, \cdots, n,$$

and

$$\pi_n u_s = u_{\pi_n s}, \quad s \in [0, \ell].$$

Following the Euler method of piecewise linear approximation, a sequence $\{u_t^n\}$ of approximations is defined by the following equation:

$$u_t^n = T_t v + \int_0^t T_{t-\pi_n s} F(\pi_n u_s^n) ds + \int_0^t T_{t-\pi_n s} \Sigma_{\pi_n s} dW_s \,. \tag{2.2}$$

For $t \in (\frac{k}{n}\ell, \frac{k+1}{n}\ell]$, it reduces to, with $T(t) = T_t$,

$$\begin{aligned}(2.3)\qquad u_t^n &= T_t v + (\frac{\ell}{n})\sum_{i=0}^{k-1} T(t-\frac{i}{n}\ell)F(u_{i\ell/n}^n) + (t-\frac{k}{n}\ell)T(t-\frac{k}{n}\ell)F(u_{k\ell/n}^n)\\ &+ \sum_{j=0}^{k-1} T(t-\frac{j}{n}\ell)\Sigma_{j\ell/n}[W_{(j+1)\ell/n} - W_{j\ell/n}] + T(t-\frac{k}{n}\ell)\Sigma_{k\ell/n}(W_t - W_{k\ell/n}).\end{aligned}$$

Let $C_\ell(H) = C([0,\ell], H)$ denote the space of H-valued continuous functions on $[0,\ell]$ with the sup-norm

$$|u.|_\ell = \sup_{0\le t\le \ell} |u_t| \ .$$

We will show that the sequence $\{u_t^n\}$ of approximate solutions converges a.s. in $C_\ell(H)$ to the solution u_t of Eq.(2.1). To this end, we need two technical lemmas.

Lemma 2.1. For any $\ell > 0$ and $p > 0$, there exists a constant $K > 0$ independent of n, such that

$$(2.4)\qquad E\{\sup_{0\le s\le \ell} |\int_0^s T(s-r)\Sigma_r dW_r|^p\} \le K$$

and

$$(2.5)\qquad E\{\sup_{0\le s\le \ell} |\int_0^s T(s-\pi_n r)\Sigma_{\pi_n r} dW_r|^p\} \le K \ .$$

(Proof). By a Burkholder type of inequality for a stochastic evolutional integral (see Remark 2 following Theorem 1 in Tubaro [5]), we have

$$(2.6)\qquad E\{\sup_{0\le s\le \ell} |\int_0^s T(s-r)\Sigma_r dW_r|^p\} \le c_1\{\int_0^\ell \|\Sigma_r\|_R^2 dr\}^{p/2}$$

for some $c_1 > 0$. By condition (A.3), the inequality (2.4) follows. To show (2.5), it suffices to note that, by a semigroup property,

$$\int_0^s T(s-\pi_n r)\Sigma_{\pi_n r} dW_r = \int_0^s T(s-r)[T(r-\pi_n r)\Sigma_{\pi_n r}]dW_r,$$

so that (2.4) implies (2.5). □

Lemma 2.2. For any $\ell > 0$, there exists a constant $K > 0$ independent of n, such that

$$\int_0^\ell \|\Sigma_s - T(s-\pi_n s)\Sigma_{\pi_n s}\|_R^2 ds \le K n^{-2\delta}.$$

where $\delta = \min\{\alpha, \nu\}$.

(Proof). Clearly we have

$$\int_0^\ell \|\Sigma_s - T(s-\pi_n s)\Sigma_{\pi_n s}\|_R^2 ds \le 2\int_0^\ell \{\|[I-T(s-\pi_n s)]\Sigma_s\|_R^2 + \|T(s-\pi_n s)(\Sigma_s - \Sigma_{\pi_n s})\|_R^2\} ds$$

$$\le c_1 \int_0^\ell \{\|[I - T(s-\pi_n s)]\Sigma_s\|_R^2 + \|\Sigma_s - \Sigma_{\pi_n s}\|_R^2\} ds \ .$$

Here and thereafter, c_i's are some generic positive constants. By (A.3), the Hölder continuity gives

$$\int_0^\ell \|\Sigma_s - \Sigma_{\pi_n s}\|_R^2 ds \le L^2 \int_0^\ell (s - \pi_n s)^{2\nu} ds \le c_2 n^{-2\nu}. \tag{2.7}$$

Furthermore,

$$\int_0^\ell \|[I - T(s-\pi_n s)]\Sigma_s\|_R^2 ds \le \int_0^\ell (s - \pi_n s)^{2\alpha} \|A^\alpha \Sigma_s\|_R^2 ds \le c_3 n^{-2\alpha} \tag{2.8}$$

which follows from an estimate on p.74 in Pazy [6] and condition (A.3). In view of (2.7) and (2.8), the lemma is proved. □

With the aid of the lemmas, we are ready to prove the first convergence theorem.

Theorem 1. Under the assumptions (A.1) to (A.3), the sequence $\{u_t^n\}$ of approximate solutions converges in $C_\ell(H)$ almost surely to the solution u_t of Eq.(2.1), that is,

$$\lim_{n\to\infty} \sup_{0\le t\le \ell} |u_t - u_t^n| = 0 \quad \text{a.s. in } \omega \ .$$

(Proof). First let us set

$$d_n(t) = \sup_{0\le s\le t} |u_s - u_s^n| \ .$$

Then, from Eqs. (2.1) and (2.2), we have

$$d_n(t) \le \xi_n(t) + \eta_n(t) \ , \tag{2.9}$$

where

$$\xi_n(t) = \sup_{0\le s\le t} \left| \int_0^s [T(s-r)F(u_r) - T(s-\pi_n r)F(\pi_n u_r^n)] dr \right| \ ,$$

$$\eta_n(t) = \sup_{0\le s\le t} \left| \int_0^s [T(s-r)\Sigma_r - T(s-\pi_n r)\Sigma_{\pi_n r}] dW_r \right| \ .$$

By Condition (A.2) and the estimate on p.74 in [6] again, we obtain

$$
\begin{aligned}
(2.10)\quad \xi_n(t) &\leq \sup_{0\leq s\leq t}\int_0^s \|[I - T(r-\pi_n r)]T(s-r)F(u_r)|dr \\
&\quad + \sup_{0\leq s\leq t}\int_0^s |T(s-\pi_n r)[F(u_r) - F(\pi_n u_r^n)]|dr \\
&\leq \sup_{0\leq s\leq t}\int_0^s c_1(r-\pi_n r)^\tau (s-r)^{-\tau}|F(u_r)|dr + \int_0^t c_1|u_r - \pi_n u_r^n|dr \\
&\leq \frac{c_2}{n^\tau}\sup_{0\leq s\leq \ell}\int_0^s (s-r)^{-\tau}dr\cdot \max_{0\leq r\leq \ell}|F(u_r)| + c_1\int_0^t |u_r - \pi_n u_r|dr \\
&\quad + c_1\int_0^t |\pi_n u_r - \pi_n u_r^n|dr \\
&\leq \theta_n + c_1\int_0^t d_n(r)dr\ ,
\end{aligned}
$$

where $0 < \tau < 1$ is any constant, and

$$
\theta_n = \frac{c_3}{n^\tau}\max_{0\leq r\leq \ell}|F(u_r)| + c_1\ell\max_{0\leq r\leq \ell}|u_r - \pi_n u_r|
$$

is a sequence of random variables. Applying (2.10) to (2.9), and further invoking the Grownwall inequality, we get

$$
d_n(\ell) \leq c_4\theta_n + c_4\eta_n(\ell)\ .
$$

Since $\theta_n \to 0$ a.s. as $n \to \infty$ by the pathwise continuity of u_t in H, there remains only to prove that $\eta_n(\ell) \to 0$ a.s.. To this end, we apply the Chebyshev inequality to get

$$
P\{\eta_n(\ell) > \frac{1}{n^\beta}\} \leq n^{p\beta}E\{\eta_n^p(\ell)\}\ , \quad \text{for any } \beta > 0\ , \tag{2.11}
$$

where, by Lemme 2.1, the right hand side is finite. Therefore, by invoking a Burkholder type of inequality as given by (2.6) and by Lemma 2.2, we get

$$
E\{\eta_n^p(\ell)\} \leq c_1\{\int_0^\ell \|\Sigma_s - T(s-\pi_n s)\Sigma_{\pi_n s}\|_R^2 ds\}^{\frac{p}{2}} \leq c_2 n^{-p\delta}\ ,
$$

so that the inequality (2.11) yields

$$
P\{\eta_n(\ell) > \frac{1}{n^\beta}\} \leq c_2 n^{-p(\delta-\beta)}\ . \tag{2.12}
$$

Since $\delta = \min\{\alpha, \nu\} > 0$, we may choose β and p satisfying $0 < \beta < \delta$ and $p > (\delta-\beta)^{-1}$, so that the series $\sum_{n=1}^\infty n^{-p(\delta-\beta)} < \infty$. By the Borel-Cantelli lemma, (2.12) implies that $\eta_n(\ell) \to 0$ a.s. as $n \to \infty$. This completes the proof of Theorem 1. □

3. Convergence in Eigenfunction Expansion

Now we turn to the question of a.s. convergence for the sequence of approximate solution obtained by an eigenfunction expansion. To this end, let A be a self-adjoint strongly elliptic operator with the orthonormal set $\{e_1, e_2, \cdots, e_n, \cdots\}$ of eigenfunctions in H. Let $\lambda_1, \lambda_2, \cdots, \lambda_n, \cdots$, denote the corresponding eigenvalues so that $\lambda_n \to \infty$ as $n \to \infty$.

Let $V_n \subset H$ denote the linear subspace spanned by $\{e_1, e_2, \cdots, e_n\}$ and let $P_n : H \to V_n$ be the orthogonal projection. By the method of eigenfunction expansion, or, more precisely, the Galerkin method, a sequence $\{u_t^{(n)}\}$ of approximate solutions is defined by the following equation:

$$u_t^{(n)} = v_n + \int_0^t [Au_s^{(n)} + F_n(u_s^{(n)})]ds + \int_0^t \Sigma_s^n dW_s , \tag{3.1}$$

where $u_t^{(n)} \in V_n$, $v_n = P_n v$, $F_n(v) = P_n F(P_n v)$ and $\Sigma_s^n = P_n \Sigma_s P_n$. The equation (3.1) can be rewritten in the evolutional integral form:

$$u_t^{(n)} = T_t v_n + \int_0^t T_{t-s} F_n(u_s^{(n)})ds + \int_0^t T_{t-s}\Sigma_s^n dW_s . \tag{3.2}$$

Denote by ζ_t^n the error of approximation given by

$$\zeta_t^n = u_t - u_t^{(n)} .$$

Then, in view of (2.1) and (3.2), we get

$$\zeta_t^n = T_t \Delta v_n + \int_0^t T_{t-s}\Delta F_n(s)ds + \int_0^t T_{t-s}\Delta\Sigma_s^n dW_s , \tag{3.3}$$

where

$$\Delta v_n = v - v_n ,$$
$$\Delta F_n(s) = F(u_s) - F_n(u_s^{(n)}) ,$$
$$\Delta\Sigma_s^n = \Sigma_s - \Sigma_s^n .$$

We will prove the a.s. convergence result under the following assumptions:

(B.1) In addition to condition (A.1), let A be self-adjoint with a complete set of eigenfunctions $\{e_n\}$ such that the corresponding eigenvalues $\lambda_n \to \infty$ as $n \to \infty$.

(B.2) Condition (A.2) holds.

(B.3) $\Sigma_\cdot : [0, \ell] \to \mathcal{L}(H_R, H)$ is continuous such that the following inequality holds:

$$\int_0^\ell \|\Sigma_s - \Sigma_s^n\|_R^2 ds \le K(\ln n)^{-(1+\varepsilon)} ,$$

for some positive constants K and ε.

The proof depends critically on the following lemma, which can be proved as Lemma 5.1 in Chow and Menaldi [7].

Lemma 3.1. Let

$$\eta_t = \int_0^t T_{t-s}\Sigma_s dW_s \ .$$

Then, for every $\ell > 0$, there exists a constant $b > 0$ such that

$$P\{\sup_{0\leq s\leq t} |\eta_s| > r\} \leq 3\exp\{-br^2/\sigma^2\}\ , \quad 0 < t \leq \ell,$$

for any $r > 0$, where

$$\sigma^2 = \int_0^\ell \|\Sigma_s\|_R^2 ds \ .$$

The second a.s. convergence theorem can be stated as follows:

Theorem 2. Let the conditions (B.1) to (B.3) be satisfied. Then the sequence of approximate solutions $\{u_t^{(n)}\}$ converges almost surely to the solution u_t in $C_\ell(H)$, that is,

$$\lim_{n\to\infty} \sup_{0\leq t\leq \ell} |u_t - u_t^{(n)}| = 0 \quad \text{a.s. in } \omega \ .$$

(Proof). Since the difference $\zeta_t^n = (u_t - u_t^{(n)})$ satisfies the equation (3.3), we have

$$|\zeta_t^n|^2 \leq 2\{|\xi_t^n|^2 + |\eta_t^n|^2\} \tag{3.4}$$

where

$$\xi_t^n = T_t\Delta v_n + \int_0^t T_{t-s}\Delta F_n(s)ds \tag{3.5}$$

and

$$\eta_t^n = \int_0^t T_{t-s}\Delta\Sigma_s^n dW_s \ . \tag{3.6}$$

Clearly,

$$|\xi_t^n|^2 \leq 2\{|T_t\Delta v_n|^2 + |\int_0^t T_{t-s}\Delta F_n(s)ds|^2\} \ . \tag{3.7}$$

By definition and the estimate on p.74 in [6] again, we obtain

$$\begin{aligned}
|\int_0^t T_{t-s}\Delta F_n(s)ds|^2 &\leq \{\int_0^t \|A^\tau T_{t-s}\|\, |A^{-\tau}\Delta F_n(s)|ds\}^2 \\
&\leq \int_0^t \|A^\tau T_{t-s}\|^2 ds \int_0^t |A^{-\tau}\Delta F_n(s)|^2 ds \\
&\leq c_1 \int_0^t |A^{-\tau}[F(u_s) - P_n F(u_s^{(n)})|^2 ds \\
&\leq c_2 \int_0^t \{|A^{-\tau}(I-P_n)F(u_s)|^2 + |F(u_s) - F(u_s^{(n)})|^2\}ds \ ,
\end{aligned}$$

where $0 < \tau < \frac{1}{2}$ is any constant. In veiw of condition (B.2) and the following estimate

$$\sup_{0\leq s\leq \ell} |A^{-\tau}(I-P_n)F(u_s)| \leq c_3\lambda_n^{-\tau} \sup_{0\leq s\leq \ell} |F(u_s)| \ ,$$

we deduce that

$$|\int_0^t T_{t-s}\Delta F_n(s)ds|^2 \leq c_4 \int_0^t |\zeta_s^n|^2 ds + c_5\lambda_n^{-2\tau} \sup_{0\leq s\leq \ell} |F(u_s)|^2 \ . \tag{3.8}$$

When the above result (3.8) is used in (3.7), which is then substituted in (3.4), we get

$$|\zeta_t^n|^2 \leq c_6 \int_0^t |\zeta_s^n|^2 ds + c_6 \sup_{0\leq s\leq t} |\eta_s^n|^2 + \theta_n \ ,$$

where

$$\theta_n = c_6|\Delta v_n|^2 + c_6\lambda_n^{-2\tau} \sup_{0\leq s\leq \ell} |F(u_s)|^2$$

is a sequence of random variables that tends to zero a.s. as $n \to \infty$. Again by Gronwall's inequality, this implies that

$$\sup_{0\leq s\leq \ell} |\zeta_s^n|^2 \leq c_7\theta_n + c_7 \sup_{0\leq s\leq \ell} |\eta_s^n|^2 \ .$$

Since $\theta_n \to 0$ a.s., there remains to prove that $\sup_{0\leq s\leq \ell} |\eta_s^n| \to 0$ a.s.. But this follows from Lemma 3.1 as to be shown. Choose a sequence $r_n = (\ln n)^{-\varepsilon/4}$. Apply Lemma 3.1 to get

$$P\{\sup_{0\leq s\leq \ell} |\eta_s^n| > r_n\} \leq 3\exp\{-b r_n^2/\sigma_n^2\} \ ,$$

where

$$\sigma_n^2 = \int_0^\ell \|\Delta\Sigma_s^n\|_R^2 ds \leq K(\ln n)^{-(1+\varepsilon)} \ ,$$

by condition (B.3). Hence

$$P\{\sup_{0\leq s\leq \ell} |\eta_s^n| > r_n\} \leq 3\exp\{-\frac{b}{K}(\ln n)^{1+\varepsilon/2}\} \ ,$$

so that, by the Borel-Cantelli lemma, the theorem is proved. □

REFERENCES

1. E. Pardoux, Equation aux dériveés partielles stochastiques non linéaires monotone, Thesis, Université Paris XI, 1975.

2. N. Ikeda and S. Watanabe, Stochastic Differential Equations and Diffusion Processes, North-Holland Mathematical Library, 1981.

3. P.L. Chow, Large deviation problem for some parabolic Itô equations, preprint (1989).

4. A. Chojnowska-Michalik, Stochastic differential equations in Hilbert spaces, Banach Center Pub. 5, 1979.

5. L. Tubaro, An estimate of Burkholder type for stochastic processes defined by the stochastic integral, Stochastic Analysis and Applications, (1984), 187-192.

6. A. Pazy, Semigroups of Linear Operators and Applications to Partial Differential Equations, Springer-Verlag, New York, 1983.

7. P.L. Chow and J.L. Menaldi, Exponential estimates in exit probability for some diffusion processes in Hilbert spaces, Stochastics and Stoch. Reports, 29 (1990), 377-393.

International Series of Numerical Mathematics, Vol. 102, © 1991 Birkhäuser Verlag Basel

TWO RESULTS ON NELSON DIFFUSIONS

Gianfausto Dell'Antonio *

Dip. di Matematica , Universita' Roma I

and S.I.S.S.A Trieste

* C.N.R, GNFM

1. INTRODUCTION

I will give here a short presentation of two results in the context of Nelson diffusions. The first is a Lyapunov-type theorem through which we estabilish conditions under which the nodes are not attained. The second proves continuity (in a suitable topology) of the map which associates to each Schroedinger evolution (defined by the potential and the initial data) the corresponding Nelson diffusion.

The results described here have been obtained in collaboration with A.Posilicano, a Ph.D. student at SISSA; in fact the results described in detail in [Po (91)] are an outgrowth of his Master Thesis.

I will only describe the strategy and outline some proofs. Details will be given elsewhere [Po (91) , D(91)].

Let ρ_t be a family of probability densities on R^d, and denote by Z_ρ the nodal set $Z_\rho=\{(t,x)\in R^{d+1} : \rho(t,x)=0\}$. Let b be a time-dependent vector field on $Z^c{}_\rho \equiv R^{d+1} - Z_\rho$ and assume that on $Z^c{}_\rho$ the Fokker-Plank equation

$$\partial\rho/\partial t = 1/2\, \Delta\rho - \mathrm{div}\,(\rho b) \tag{1.1}$$

is satisfied in a weak sense.

A weak solution , if it exists, of the stochastic differential equation

$$X_t = X_0 + \int_0^t b(s,X_s)ds + B_t \tag{1.2}$$

will be called a Nelson diffusion.

A particular class of Nelson diffusions are the Schroedinger diffusions, which are such that

$$\rho_t=|\psi_t|^2 \ , \quad b_t = (\mathrm{Re}+\mathrm{Im})\nabla\psi_t \tag{1.3}$$

where ψ_t is a solution of the Schroedinger equation

$$i\partial\psi/\partial t = 1/2\, \Delta\psi + V\psi \tag{1.4}$$

(regularity assumptions on the potential V will be specified later).

An unpleasant property of these diffusions is that their drifts are too singular to be handled by traditional approaches. The problem of existence of Schroedinger processes was solved by Carlen in [C(84)] for potentials form-bounded by $\Delta/2$ and initial data in $H^1(R^d)$. His existence theorem provides a Borel probabilty measure on $\Omega \equiv C(R^+,R^d)$ for which the stochastic process X_t defined by $X_t(\xi)=\xi(t)$ is a Markov process with density $|\psi_t|^2$ and is a weak solution of (1.2) with drift given by (1.3).

Carlen's theorem provides a map from a space of solutions of the Schroedinger equation to the space $M_1(\Omega)$ of probabilty measures on Ω .

It is then natural to consider the following continuity problem.

Let $\{\psi_t^n\}$, $n\geq 1$ be a sequence of solutions of Schroedinger equations with potentials V_n and initial data ψ_0^n , and let $\{P_n\}$ be the corresponding sequence of probabilty measures given by Carlen's theorem.

Which convergence of the V_n and ψ_0^n will imply that the P_n converge?

We will give an answer to this question in Section 3.

In Section 2 we will discuss instead a problem which is directly related to the possibility to provide a probabilistic construction of Nelson diffusions [Cr (87)], [Z(86)] (the proof given by Carlen proceeds by constructing the transition kernels and uses methods of Functional Analysis).

Using Lyapunov functions we prove that the nodal set is not attainable If one assumes that the drift b is sufficiently regular in every bounded open set disjoint from the nodal set . Notice that this is true for Schroedinger diffusions if ψ_t is in $H^1(R^d)$ for each t and $|\psi|_1$ is bounded on every bounded interval of R^+ ; this is in particular implied by the finite energy condition used by Carlen.

2. THE NODAL SET CANNOT BE ATTAINED

Assume that the drift b is sufficiently regular on $Z^c{}_\rho$. From standard existence theorems for s.d.e. there exists a local Markovian diffusion in every bounded domain D in $Z^c{}_\rho$ with regular boundary ∂D. The process will be defined up to a stopping time t_D ; if t_D does not increase without bound for a sequence D_n which invades $Z^c{}_\rho$, the process will hit the nodes or escape to infinity in finite time. For a Nelson process, $\rho(t,x)$ represents the probability density to find the "particle" at the point x at time t.

Since $\rho(t,x)$ vanishes on the nodal set and at infinity, it is very plausible that in some sense the process has a "self-regulating" mechanism which prevents it from reaching the nodes or escaping to infinity in a finite time. Notice that Nelson diffusions are in some sense self-regulating since the drift b and the density ρ must at each time satisfy the Fokker-Planck equation (1.1) ; for Schroedinger diffusions this condition is similar to conservation of energy for the motion of a classical particle in the potential V .

Here we shall consider only the problem of avoiding the nodes.The first guess one can make about the regulating mechanism is that the drift field becomes indeed very large when one approaches the nodal set, but it points away from this set, preventing the trajectory from reaching it.

For a small class of Schroedinger diffusions this guess is indeed correct. For example, if the process corresponds to an eigenfunction of the Schroedinger equation (so that the process is stationary) under suitable regularity assumptions one can prove [Po(91)], that in a sufficiently small neighborhood of Z_ρ the distance function from Z_ρ increases under the classical flow induced by the drift field b .

One can then use a standard Lyapunov theorem, using $\ln(d(Z_\rho))$ as Lyapunov function, to prove that with probabilty one the nodal set cannot be attained In the non-stationary case, this first guess is in general wrong. This can be seen from the following simple example [Po(91)] .

Let ψ_t be the solution of the Schroedinger equation for the harmonic oscillator in one dimension with initial condition

$$\psi_0 = \lambda \exp\{-x^2/2\} + \mu\, x \exp\{-x^2/2\}\ , \quad 2\lambda^2 + \mu^2 = 2\ ,$$

One verifies that

$$Z_\rho = \{(t,x) \in R^2 : x = \pm\lambda/\mu\ ,\ t = k\pi\ ,\ k = 1,2,..\}$$

and in a neighborhood of a nodal set

$$b(t,x) = \mu^2(-x\pm\lambda/\mu)+\lambda\mu(t-k\pi)/[\mu^2(-x\pm\lambda/\mu)^2+\lambda^2(t-k\pi)^2] + O(1)$$

Therefore at each point of the nodal set there are cones in which the distance function $[(t-k\pi)^2 + (-x\pm\lambda/\mu)^2]^{1/2}$ decreases under the flow of b ; one could say pictorially that the "regulating mechanism" acts in a more sophisticated way, namely in these cones the "particle" is drawn very fast to Z_ρ , but it is then ejected immediately in the opposite direction and pushed quickly far from Z_ρ , so that indeed the probabilty to find the particle in a small neighborhood N vanishes with the radius of N .

There is therefore no hope to use a 'pathwise" Lyapunov theorem.

The key fact to find a solution is the remark that for Nelson diffusions one can give a-priori estimates on the density of the process in terms of the given function ρ. In particular one has

Lemma 2.1 [Po(91)]

Let ρ_t be a family of probability densities on R^d, let b be a time-dependent vector field on Z^c_ρ such that $b \in L^\infty_{loc}(R^+, L^p_{loc}(Z^c_\rho))$, p>d, and divb is a measurable function. Suppose moreover that ρ and b satisfy the Fokker-Plank equation (1.1) in a weak sense on every bounded open domain E in Z^c_ρ . Let X_t be the evaluation process and denote by σ the first exit time of X_t from E . Then for every bounded non-negative ρ-measurable function f one has

$$E^\sigma{}_x\int_0^{\sigma\Lambda t} f(s,X_s)dt \le \int_E f(s,y)\rho(s,y)dyds \quad \forall x\in E$$

where $E^\sigma{}_x$ denotes expectation w.r.to $P^\sigma{}_x$, the weak solution of the s.d.e.

$$X_t = x + \int_0^t b(s,X_s)ds + B^\sigma{}_t$$

($P^\sigma{}_x$ exists by L^p-integrability of b [Ca(89)],[P(75)]) /

This result allows to prove a Lyapunov-type theorem in which the pathwise conditions are replaced by conditions in the mean.
In particular, using as Lyapunov function the Stein regularized distance, one can prove [Po(91)]

2.2 Theorem

Let ρ_t and b as in Lemma 2.1. Suppose that Z_ρ is a closed set and

$$\rho d^{-2}(Z_\rho) \in L^1{}_{loc}(U) \quad , \quad \rho|b|d^{-1}(Z_\rho) \in L^1{}_{loc}(U)$$

where U is a neighborhood of Z_ρ. Then for every bounded open set $D\in R^+\times R^d$ with regular boundary ∂D the corresponding Nelson diffusion on D stopped at ∂D does not reach the nodes of ρ with probabilty one . /

2.3 Remark

If one assumes more regularity of ρ , e.g. $\rho\in C^{1,2}(Z^c{}_\rho)$, one can use as Lyapunov function the distance from the nodal set. In this case one reproduces the results in [B(87)] (these Authors do not place their analysis in the context of stability theory).

3 CONTINUITY OF THE CARLEN MAP

As remarked in the introduction, Carlen [Ca(84)] has constructed a (Nelson) Process in R^d with density $\rho(t,x)$ and drift field $b(t,x)$ under the condition that ρ and b satisfy a weak form of the continuity equation and a "finite energy condition" . More precisely let

$$v=b^{-1/2}\log\rho \text{ on } Z^c{}_\rho\,, \quad u=\nabla\log\rho \text{ on } Z^c{}_\rho\,, \quad u=v=0 \text{ on } Z_\rho \qquad (3.1)$$

and assume that

A) $\forall f\in C^1{}_b(R^d)$ the function $t\to\int f(x)\rho(t,x)dx$ is differentiable and

$$d/dt\int f(t,x)\rho(t,x)dx = \int v(t,x)f(t,x)\rho(t,x)dx$$

B) $\int_0^T \int(||u||^2+||v||^2)\rho(t,x)dxdt < +\infty \quad \forall\, T>0$

Then Carlen proves that there exists a Borel probabilty measure P on $C(R^+,R^d)$ such that

1) The evaluation process X_t is a square-integrable Markov process under P, with jointly measurable version
2) The distribution of X_t has density ρ
3) $B_t \equiv X_t-X_0-\int_0^t b(s,X_s)ds$ is a brownian motion under P, with the natural filtration.

We shall use the notation $X_C(v,\rho)$ for the process constructed by Carlen, and call $\{v,\rho\}\Rightarrow X_C(v,\rho)$ the Carlen map.

The process so constructed is a Schroedinger process for a potential V if $\{u,v\}$ provide a weak solution in $Z^c{}_\rho$ of the following "Madelung" equation

$$\partial u/\partial t = -\nabla V + u.\nabla u - v.\nabla v + \Delta u \qquad (3.2)$$

and moreover there is a function S such that $v=\nabla S$ weakly.

Setting $R=1/2\log\rho$ and $\psi=\exp\{R+iS\}$, the following identity is then satisfied weakly in $Z^c{}_\rho$, at least if V is sufficiently regular

$$i\partial\psi/\partial t = -1/2\ \Delta\psi + V\psi \qquad \psi_0(x) = \exp\{R(0,x)+iS(0,x)\}$$

When this is the case, we shall use the notation $X_S(\psi_0,V)$ for Carlen's map.

One can prove [D(91)]] that Carlen's theorem provides a Schroedinger process (so that $X_S(\psi_0,V)$ is well defined) for every potential which is form-bounded with respect to $-1/2\Delta$ with relative bound smaller than one, and for every initial datum ψ_0 in $H^1(R^d)$.

To see this, let K be the self-adjoint represenation of $-1/2\Delta$ on $L^2(R^d)$, and V be a real-valued measurable function on R^d.

Recall that V , acting by pointwise multiplication , is K-form bounded, with bound smaller than one, if one can find $a\in[0,1)$ and $b\geq 0$ such that

$$|(\psi,V\psi)| \leq a(\psi,\psi) + b(\psi,\psi) \quad \forall\ \psi\in H^1(R^d) \qquad (3.3)$$

where (.,.) is the scalar product in R^d.

Let H be the unique self-adjoint operator associated to the sum of the quadratic forms of K and V [R(75)]. Then $t\rightarrow\exp\{-itH\}$ is a continuous one-parameter group of bounded invertible linear operators on $H^1(R^d)$ and

$$||\exp\{-itH\}||_1 \leq (2(b+1)+a)(1-a)^{-1}$$

One can then prove ([D(91)],Th 2.1) that , under the assumption

$$\psi_0\in H^1(R^d) \qquad (3.4)$$

there are unique jointly measurable functions $\psi(t,x)$ and $\nabla\psi(t,x)$ such that

$\psi(t,x)=[\exp\{-itH\}\psi_0](x)$, $\nabla\psi(t,x)=\nabla[\exp\{-itH\}\psi_0](x)$

Moreover, defining

$$\rho \equiv \psi_0{}^*\psi_0\ ; \quad u = \mathrm{Re}\nabla\log\psi\ ,\ v = \mathrm{Im}\nabla\log\psi \text{ in } Z^c{}_\rho\ ; \quad u = v = 0 \text{ in } Z_\rho$$

conditions A and B of Carlen's theorem are satisfied.

3.1 Remark

Under the finite energy condition B , one has

$$E_P\int_0^T ||b(s,X_s)||^2 ds < +\infty \quad \forall\ T \geq 0$$

so that

$$P|_{FT} < W|_{FT}\ , \quad H_{FT}(P;W) = 1/2\ E_P\int_0^T ||b(s,X_s)||^2 ds$$

where $W = \int W_x \rho(0,x)dx$, W_x is standard Wiener measure supported on the space of continuous paths in R^d starting at x, and $H_{FT}(P;W)$ is relative entropy.

Therefore condition B is a relative entropy condition. In the case of Schroedinger diffusions it is called finite energy condition since it may be written as [Ca(84)]

$$\int_0^T ||\nabla\psi_t||^2 dt < +\infty \qquad /$$

With these definitions, it it natural to inquire whether the Carlen maps $X_C(v,\rho)$ and $X_S(\psi_0,V)$ are continuous. The answer will of course depend on the topologies one chooses for the processes, for the sets $\{\rho,v\}$ of data satifying A and B, and for the the sets $\{\psi_0,V\}$ satisfying (3.3) and (3.4).

For the map X_C one can remark that Carlen's proof is obtained by proving weak convergence to $X_C(\rho,v)$ of a suitable sequence $X_C(v_n,\rho_n)$ where each ρ_n is smooth and strictly positive on every bounded set in R^d, while v_n is bounded on bounded sets. The existence of $X_C(v_n,\rho_n)$ follows then from standard results in the theory of Stochastic Differential Equations.

The sequence $\{ v_n,\rho_n\}$ converges to $\{\rho,v\}$ in the following sense

$$\int |v_n-v|\rho dx \to 0 \ , \ \int (\sqrt{\rho_n}-\sqrt{\rho})^2 dx \to 0 \qquad (3.5)$$

It is then natural to conjecture that the following continuity result holds:

3.2 Conjecture [G(86)]

A sequence of (finite-energy) Carlen processes $X_C(v_n,\rho_n)$ converges weakly to the process $X_C(\rho,v)$ if

$$||v_n\sqrt{\rho_n}-v\sqrt{\rho}|| \to 0 \ , \ ||\sqrt{\rho_n}-\sqrt{\rho}|| \to 0 \qquad (3.6)$$

(where $||.||$ is the norm in $L^2(R^d)$) , i.e. the Carlen map X_C is continuous in the weak topology of Nelson processes if one endows the data $\{\rho,v\}$ with the topology given by (3.6). /

It is not known at present whether this conjecture is true. One should remark that the first condition in (3.6) can be written, at least formally,

$$||(\sqrt{\rho}/\sqrt{\rho_n})\ v_n\sqrt{\rho_n} - v\sqrt{\rho}|| \Rightarrow 0$$

In Carlen's theorem the sequence $\{ v_n,\rho_n \}$ was chosen so that $1/\rho_n$ is for each n uniformly bounded on bounded sets and ρ/ρ_n converges to one pointwise uniformly on bounded sets. For a generic sequence ρ_n which converges to ρ in the sense of (3.6), the function ρ/ρ_n is unbounded and defined only outside the nodal set of ρ_n ; under these conditions, a sufficient convergence of ρ/ρ_n to 1 is problematic, so that in general (3.5) does not follow from (3.6).

3.3 Remark.

For Schroedinger Diffusions, we have remarked that under suitable regularity assumptions, if $\{v,\rho\}$ satisfy the Madelung eqautions for the potential V, then one can find ψ_0 such that $X_C(v,\rho) = X_S(\psi_0,V)$.

Little is known in general about existence and uniqueness of the solutions of the Madelung equation for a given potential V . If V is form-bounded by K, as remarked above a solution exists for each ψ_0 in $H^1(R^d)$ but uniqueness is not known. Therefore even if $\{v_n,\rho_n\}$ is a Cauchy sequence in the topology (3.5) satisfying the Madelung equations and converging to v,ρ and moreover $X_C(v_n,\rho_n)$ converges weakly to $X_C(v,\rho)$, it is not known in general if one can find ψ_0 , V such that $X_C(v,\rho) = X_S(\psi_0,V)$.

Notice however that if V_n=V for all n , with V form-bounded by K , the conclusion is true if one passes to a subsequence ; one can disintegrate with respect to the initial distribution and prove convergence of $\psi_{o,n}$ in $L^2(R^d)$ through a subsequence, using the compact embedding of H^1 into L^2. /

We shall concentrate here on the problem of the continuity of the Carlen map $X_C(\psi_0,V)$ with respect to a suitable topology in the space of data $\{\psi_0,V\}$.

Consider either the Kato-class K_d of potentials with Kato norm $||V||_K$ or, for d=3, the Banach space R of Rollnick potentials with norm $||V||_R$.

The following holds [D(91)]

3.4 Theorem

Let V, $V_n \in K_d$, n≥1, $\psi_0,\psi_0{}^n \in H^1(R^d)$ n≥1. If P, P_n are the probability measures on $C(R^+,R^d)$ which correspond to the processes $X_S(\psi_0{}^n,V_n)$, $X_S(\psi_0,V)$, and if

$$\lim_{n\to\infty} || \psi_0{}^n - \psi_0 ||_1 = 0 \quad , \lim_{n\to\infty} ||V_n - V||_K = 0$$

then there exists a subsequence $\{P'_j\}$, j=1,2,.., such that

$$\lim_{j\to\infty} |||P'_j - P||| = 0$$

where $|||.|||$ is the total variation norm of a signed measure.

The same result holds when d=3 if one assumes that V, V_n are in the Banach space R and V_n converge to V in Rollnik norm. /

3.5 Remark

The convergence proved in Theorem 3.1 is stronger than weak convergence of the corresponding processes. /

3.6 Remark

It may appear that convergence of the initial data in H^1 be an unnecessarely strong asssumption ; since one can disintegrate w.r. to the initial distribution, one may expect that L^2 convergence be sufficient. However, since by assumption all potentials are in K and $||V_n - V||_K \to 0$, one has

$$\sup_n ||\exp\{-itH_n\}||_1 < C$$

$$\lim_{n\to\infty} \sup_{o\le t\le T} ||\exp\{-itH_n\}-\exp\{-itH\})||_1 = 0$$

so that

$$||\psi_0{}^n - \psi^0||_1 \to 0 \quad \Leftrightarrow \quad \psi_t{}^n \to \psi_t \text{ in energy norm} \qquad /$$

Theorem 3.4 is proved in [D(91)] as a consequence of another Theorem in which the conclusions of Theorem 3.4 about convergence of probability mesaures are derived from uniform convergence in H^1 of the sequence $\psi_t{}^n$. More precisely one has

3.7 Theorem

Let V_n,V be form bounded with respect to K with bounds smaller than one. Let H=K+V, H_n=K+V_n the corresponding operators defined from quadratic forms. Consider a sequence of initial data $\psi_o{}^n$ in $H^1(R^d)$ and let $\psi_t{}^n$ be the sequence defined by $\psi_t{}^n=\exp\{-itH_n\}\psi_o{}^n$. Denote by $\{P_n\}$ the sequence of probability measures on $C(R^+,R^d)$ associated to $X_S(\psi_o{}^n, V_n)$. If

$$\lim_{n\to\infty}\sup_{0\le t\le T}\| \psi_t{}^n - \psi_t\|_1 = 0 \quad \forall\ T\ge 0$$

where $\psi_t = \exp\{-itH\}\psi_o$, then there exists a subsequence $\{P'_j\}$ such that

$$\lim_{j\to\infty}|||\ P'_j - P\ ||| = 0 \qquad /$$

The proof of Theorem 3.7 is obtained by considering first the sequence of processes $X_S(\psi_o{}^n,V_n)$ each stopped at a finite distance from its nodal set, and proving convergence of the measures associated to the stopped processes. Convergence of the measures associated to the sequence $X_S(\psi_o{}^n,V_n)$ follows then from Lemma 3.8 below., which is similar to Lemma 11.1.1 in [S(79)].

3.8 Lemma

Let P , P_n n≥1, be probability measures on $\Omega \equiv C(R^+,R^d)$, which have a disintegration with respect to X_0 ; let $\mu=P.X_0^{-1}$, $\mu_n=P_n.\ X_0^{-1}$. Let $\{\tau_k\}$ k≥1 be a non-decreasing sequence of F_t stopping times, where $\{F_t\}$ is the natural filtration. Let $\{P_x{}^k\}$, $\{P_x{}^{n,k}\}$ be sequences of probability measures on Ω such that for each k≥1 $P_x{}^k\,|F^k = P_x|\,F^k$, $P_x{}^{n,k}\,|F^k = P_x{}^n\,|F^k$,with $F^k=\{E\in F : E\cap\{\tau_k\le t\}\in F_t\ \forall\ t\ge 0\}$.

Assume that for each $k \geq 1$ there exist subsequences $(P'_x{}^j\}$, $\{\mu'^j\}$ (depending on k) such that

$$\lim_{j\to\infty} \int_{R^d} \| P'_x{}^j - P_x{}^k \| \, d\mu'^j = 0 \qquad (C)$$

If

$$\lim_{k\to\infty} P(\tau_k < t) = o \ \forall t>0 \ \text{ and } \ \lim_{n\to\infty} \|\mu_n - \mu\| = 0$$

then there exists a subsequence $\{P''_j\}$ such that

$$\lim_{j\to\infty} ||| P''_j - P ||| = 0 \qquad /$$

To verify condition C in Lemma 3.8, one needs good pointwise estimates on the behaviour of functions in the Banach space

$$W(T) = C([0,T], H^1(R^d)) \ , \ \|u\|_{W(T)} = \sup_{0\leq t\leq T} \|u_t\|_1$$

which is the natural space for solutions of the Schroedinger equation with initial data in $H^1(R^d)$ and potentials which are form-bounded with respect to K. We control the local behaviour of functions in W(T) by means of a suitable parabolic capacity which is introduced and discussed in [D(91)], along the lines of a general procedure described in [A(56)]. This allows to find a good sequence of stopping times adapted to to a decreasing sequence of neighborhoods of the nodal sets of the densities ρ_n.

We refer the interested reader to [D(91)] for the details of the constructions and for complete proofs.

REFERENCES

A (56) Aronszajn.N, Smith.K. Functional Spaces and Functional Completion Ann. Inst. Fourier 6 125-185 (1955/56)

B (87) Blanchard.P , Golin.S. Diffusion Processes with singular drift fields , Comm. Math. phys 109 , 421-435 (1987)

Ca (84) Carlen.E Conservative Diffusions Comm. Math. Phys. 94 , 293-315 (1984)

Ca (89) Carlen.E , Kree.P L^p-estimates for iterated stochastic integrals BIBOS preprint N0 979 , 1989

Cr (87) Carmona.R Probabilistic Construction of Nelson Diffusions in K.Ito , N.Ikeda eds., Probabilistic Methods in Mathematical Physics , Katata 1985 Academic Press 1987

D (91) Dell'Antonio.G.F. , Posilicano.A Convergence of Nelson Diffusions Preprint 1991 , submitted to Comm. Math. Phys.

G (85) Guerra.E Carlen Processes:a new type of diffusions with singular drift in "Quantum Probability and Applications, ed. L.Accardi, W von Waldenfens, Lect. Notes in Math. 1136, Springer Verlag 1985

P (75) Portenko.N. Diffusion Processes with unbounded drift coefficients Theory Prob. and Applications 20, 27-37 (1975)

Po (91) Posilicano.A A Lyapunov type theorem for Nelson Diffusions Preprint 1991 ,submitted to Probability Theory and Applications

R (75) Reed.M. , Simon.B. Methods of Mathematical Physics vol2 Academic Press 1975

S (79) Stroock,D. Varadhan.S.R.S. Multidimensional Diffusion Processes Springer Verlag 1979

Z (86) Zheng.W. , Meyer.P. Sur la construction de certain diffusions in J.Arema , M.Yor eds. , Seminaire de Probabilite' XX (84/85) Springer Lect.Notes in Math. 1204, 334-337 (1986)

Bulk Conductivity of the Square Lattice for Complex Volume Fraction

Kenneth Golden*

Department of Mathematics
Princeton University
Princeton, NJ 08544 USA

Abstract. The bulk conductivity $\sigma^*(p)$ of the bond lattice in $\mathbb{Z}^d$ is considered, where the conductivity of the bonds is 1 with probability p or $\epsilon \geq 0$ with probability $1-p$. The behavior of $\sigma^*(p)$ for complex p is investigated numerically and analytically. We find numerical evidence that the phase transition occurring at the percolation threshold p_c is characterized by a domain "splitting," where $\sigma^*(p) = 0$ in a region containing $(0, p_c]$ while $\sigma^*(p) \neq 0$ in a region containing $(p_c, 1]$. Furthermore, the loss of analyticity at p_c does not appear to arise from a "pinching" of zeros around p_c, as in the Lee-Yang picture of phase transitions. A partition function for σ^* is introduced and is similarly analyzed.

1. Introduction.

Random resistor networks [**1-3**] based on the percolation model provide an excellent setting in which to study disordered conductors. In particular, we consider the bulk conductivity $\sigma^*(p)$ of the bond lattice in $\mathbb{Z}^d$, where the conductivity of the bonds is either 1 with probability p or $\epsilon \geq 0$ with probability $1 - p$. What is particularly interesting about this model is that for $\epsilon = 0$ it exhibits a phase transition at a critical probability $p_c \in (0,1)$, i.e. for $p < p_c, \sigma^*(p) = 0$, while $\sigma^*(p) > 0$ for $p > p_c$. This critical probability p_c coincides [**4,5**] with the percolation threshold p_c, below which $P_\infty(p) = 0$, where $P_\infty(p)$ is the probability that the origin belongs to an infinite cluster of occupied (conductivity 1) bonds, and above which $P_\infty(p) > 0$. As $p \to p_c^+$, it is believed that $\sigma^*(p)$ exhibits critical scaling, $\sigma^*(p) \sim (p - p_c)^t$, where t is called the conductivity critical exponent.

It is interesting both mathematically and physically to understand the exact nature of the conducting phase transition at p_c. In previous work [**6**] we proved that for every $\epsilon > 0$, $\sigma^*(p)$ is analytic in an open neighborhood of [0,1] in the complex p-plane. This result establishes that for $\epsilon > 0$, no phase transition can occur. As $\epsilon \to 0$, $\sigma^*(p)$ loses analyticity, at least at $p = p_c$. Analogously, with $\epsilon = 0$, the finite volume version $\sigma_L^*(p)$ of $\sigma^*(p)$, where L is the length of a side of a box of volume L^d, is a polynomial in p, which is an (entire) analytic function of p. As the infinite volume limit is taken,

*Supported in part by NSF Grant DMS - 8801673 and AFOSR Grant AFOSR-90-0203.

analyticity is again lost, at least at $p = p_c$. The question arises as to exactly how this analyticity is lost.

In the Lee-Yang picture of phase transitions [see, e.g., ref. **7**], the finite volume partition function is a polynomial in the fugacity, whose zeros lie off the positive real axis, reflecting the absence of a phase transition for finite volumes. As the infinite volume limit is taken, these zeros coalesce into a curve, and "pinch" the real axis, causing a loss of analyticity at the critical fugacity. For certain systems, even for small volumes the zeros lie on a well defined curve.

In this work we investigate $\sigma_L^*(p)$ for complex p in the case of the $d = 2$ square lattice. If the lattice has N bonds, then $\sigma_L^*(p)$ is an Nth order polynomial in p. We use the computer to calculate the conductivities of realizations of the random network, which yields the coefficients of the polynomial. Then the program Mathematica is used to obtain various graphs of $\sigma^*(p)$ in the complex p-plane. We have only considered lattices of size up to 9×9, with $\epsilon = 0.001$, but there are certain discernible features of $\sigma^*(p)$ which are common to all the graphs. The principal feature is that there appears to be a region W, which can be very roughly described by

$$W \approx \{ p \,|\, 0 \leq \text{Re } p < p_c, -0.1 < \text{Im } p < 0.1 \},$$

such that $\sigma^*(p)$ appears to vanish in W. Apparently the domain of analyticity of $\sigma^*(p)$ around [0,1] appears to be split around p_c (although it is not rigorously known that $\sigma^*(p)$ is analytic anywhere). This splitting of the domain (or the splitting of the function into two separate analytic pieces) is part of the Lee-Yang picture, where the line of zeros splits the domain of analyticity of the partition function. However, we have as yet found no evidence of zeros pinching the Re p axis around p_c. In addition to $\sigma_L^*(p)$, we shall define a partition function for the problem, and note that it has a similar behavior to $\sigma_L^*(p)$ for complex p.

2. Formulation.

We formulate the bond conductivity problem for an arbitrary graph. Let G be a finite graph consisting of N bonds $\{b_i\}$ and M vertices $\{x_i\}$. Assigned to G are N independent random variables c_i, $1 \leq i \leq N$, the bond conductivities, which take the values 1 with probability p and $\epsilon \geq 0$ with probability $q = 1 - p$. Distinguish two vertices, say $x_1 = x$ and $x_M = y$, and connect them to a battery which keeps the voltage drop between them equal to 1. The effective conductivity $\widetilde{\sigma}(\omega)$ of the network for any realization ω of the bond conductivities is just the total current $i(\omega)$ that flows through the network, which is obtained via Kirchoff's laws. We define $\sigma(p) = < \widetilde{\sigma}(\omega) >$, where the expectation $< \cdot >$ is over all 2^N realizations. For example, a two bond network has

$$\sigma(p) = p^2 \widetilde{\sigma}(1,1) + pq(\widetilde{\sigma}(1,\epsilon) + \widetilde{\sigma}(\epsilon,1)) + q^2 \widetilde{\sigma}(\epsilon,\epsilon) \ ,$$

where $\widetilde{\sigma}(1,1) = \widetilde{\sigma}(\omega)$ with $\omega = (1,1)$, and so on. For N bonds, $\sigma(p)$ is an Nth order homogeneous polynomial in p and q,

$$\sigma(p) = \sum_{k=0}^{N} \alpha_k p^{N-k} q^k$$
$$\alpha_k = \sum_{\omega^k \in \Omega^k} \widetilde{\sigma}(\omega^k), \quad q = 1 - p, \tag{2.1}$$

where $\Omega^k = \{\omega^k = (\omega_1, \ldots, \omega_N) | \omega_\ell = \epsilon \text{ for exactly k of the } \omega_\ell\text{'s}\}$.

The cases of most interest are when G is a square, cubic, or hypercubic lattice. Then, with $d = 2$ for simplicity, we take an $L \times L$ sample of the lattice and attach a perfectly conducting bus bar to each of two opposite edges of the sample. This can be accomplished [4] in the above language by attaching to each vertex of these opposing edges a perfectly conducting bond. All of these bonds from one edge meet at a new vertex x and all the bonds from the other edge meet at a new vertex y. Then x and y are connected again with the unit battery. Random bond conductivities are assigned only to the bonds in the original $L \times L$ sample. Let $\sigma_L(p)$ denote (2.1) for the effective conductivity measured between x and y. Then for $d \geq 1$, the finite volume bulk conductivity $\sigma_L^*(p)$ is defined as

$$\sigma_L^*(p) = L^{2-d} \sigma_L(p) . \tag{2.2}$$

Finally, we define the bulk conductvity $\sigma^*(p)$ by

$$\sigma^*(p) = \lim_{L \to \infty} \sigma_L^*(p). \tag{2.3}$$

For $\epsilon > 0$, the infinite volume limit in (2.3) has been shown to exist [**8-11**], and for $\epsilon = 0$ the existence of σ^* has recently been proven in the continuum [**12**].

3. Analyticity for $\epsilon > 0$.

For completeness, we state and prove here the analyticity result which motivates the present investigation. The analysis is based on an integral representation which was proved for two component stationary random media in [**10**] (see also [**13**]). The formulation there is in the continuum, but applies in the present context by replacing the continuum equations for the electric and current fields with their discrete analogs, i.e., Kirchoff's laws. We repeat here only the relevant features.

Let $s = 1/(1-\epsilon)$. We shall consider s to be a complex variable. It can be shown that $\sigma^*(p,s)$ is analytic everywhere in the s-plane except for the interval $[0,1]$. Furthermore, $\sigma^*(p,s)$ maps the upper half plane to the upper half plane, i.e., Im $\sigma^*(p,s) > 0$ when Im $s > 0$. As a consequence of these analytic properties, $\sigma^*(p,s)$ has the following integral representation,

$$1 - \sigma^*(p,s) = \int_0^1 \frac{d\mu(x)}{s-x} , \tag{3.1}$$

where μ is a positive Borel measure on $[0,1]$ which depends on p. Notice that this representation separates the dependence of $\sigma^*(p,s)$ on s from its dependence on p. (In fact, (3.1) applies even when $\epsilon = 0$.) The dependence of μ on p is most easily obtained through its moments, as follows. For $|s| > 1$, (3.1) can be expanded about a homogeneous medium ($s = \infty$ or $\epsilon = 1$), yielding

$$1 - \sigma^*(p,s) = \frac{\mu_0(p)}{s} + \frac{\mu_1(p)}{s^2} + \frac{\mu_2(p)}{s^3} + \dots , \tag{3.2}$$

$$\mu_n = \int_0^1 x^n d\mu(x) . \tag{3.3}$$

By equating (3.2) to a similar expansion of a resolvent representation for σ^*, one can obtain a formula for $\mu_n(p)$ in terms of the iterates of a self adjoint operator on L^2 ($\Omega =$ set of realizations of the bond conductivities) involving the Green's function of the discrete Laplacian. Because the bond conductivities are independent, these moments can be computed in principle, but they become very complicated. The first two are

$$\begin{aligned} \mu_0(p) &= 1 - p \\ \mu_1(p) &= \frac{p(1-p)}{d} . \end{aligned} \tag{3.4}$$

In general, $\mu_n(p)$ is an $(n+1)$-order polynomial in p.

We are now ready to state

THEOREM 3.1: *($d \geq 1$ bond problem) For every $\epsilon > 0$, there exists an open neighborhood V_ϵ in the complex p-plane such that $[0,1] \subset V_\epsilon$ and $\sigma^*(p)$ is analytic in V_ϵ.*

Proof. Fix $s = 1/(1-\epsilon) > 1$. The idea is to produce a neighborhood containing $[0,1]$ in which (3.2) converges uniformly. Since for $p \in [0,1]$, $\mu_0(p) = 1-p$ and $\mu_n(p) \geq \mu_{n+1}(p)$ for all n (via (3.3)),

$$\mu_n(p) \leq 1, \quad p \in [0,1] . \tag{3.5}$$

Now we must extend what we can of (3.5) into the complex plane. Consider $E = \{p \in \mathbb{C} | p \notin [0,1]\}$. Conformally map E onto the unit disk D in the z-plane, so that $p = \infty$ gets mapped to $z = 0$, and $[0,1]$ gets mapped to the unit circle $|z| = 1$. Let $m = n+1$. Since $\mu_n(p)$ is an mth order polynomial in p, $\mu_n(z)$ has at worst an mth order pole at $z = 0$. Thus $z^m \mu_n(z)$ is analytic in D. Since $|\mu_n(z)| \leq 1$ for $|z| = 1$, by the maximum modulus principle,

$$|\mu_n(z)| \leq \frac{1}{|z|^m} , \quad z \in D . \tag{3.6}$$

For any small $\delta' > 0$, there is a small $\delta > \delta' > 0$ such that in the annulus $A_{\delta'}$ defined by $1 \geq |z| > 1 - \delta'$

$$|\mu_n(z)| \leq (1+\delta)^m , \quad z \in A_{\delta'} . \tag{3.7}$$

For our given $s > 1$ (or $\epsilon > 0$), we can choose δ and δ' such that

$$|\mu_n(p)| \leq (1+\delta)^m < s^m , \quad p \in V_\epsilon , \tag{3.8}$$

where V_ϵ conformally maps to $A_{\delta'}$. Then (3.2) converges uniformly in V_ϵ, which proves the theorem.

Remarks. Theorem 3.1 and its proof hold for a large class of continuum systems as well, namely infinitely interchangeable media, which have recently been introduced by O. Bruno [14]. This class is a generalization of Miller's cell materials [15], where all of space is divided up into cells, such as spheres of all sizes, and then the conductivity of each cell is a random variable (independent from the others) taking two (or more) values with probability p and $1-p$. While the integral representation (3.1) holds in great generality, along with (3.5), what is needed to make the proof go through is that the $\mu_n(p)$ are polynomials in p. The proof of this fact for infinitely interchangeable media is contained in [16] (along with rigorous upper and lower bounds on $\sigma^*(p)$ for the $d=2$ bond problem with $\epsilon > 0$). We also note that Theorem 3.1, and its generalization to infinitely interchangeable media, provides a rigorous basis for the volume fraction expansions of $\sigma^*(p)$, which have been widely used for a long time [16]. Finally, Theorem 3.1 presumably does not hold for all composite media. For example, $\sigma^*(p)$ for a periodic array of spheres of volume fraction p embedded in a host material is believed to be analytic at $p=0$ only in the variable $p^{\frac{1}{3}}$, so that $\sigma^*(p)$ has a branch cut there (see, e.g., [17]).

4. Calculation of $\sigma_L^*(p)$.

In order to investigate the behavior of $\sigma_L^*(p)$ for complex p, we numerically calculate the α_k in (2.1), which depend on the effective conductivities of realizations of the lattice with random bond conductivities. It is first useful to describe a beautiful formula which gives the effective conductivity of any graph with any set of conductivities on the bonds.

Let G be a graph with M vertices and N bonds. Without loss of generality we can assume that any two vertices are joined by only one bond, since if there are n bonds joining two vertices, they can be replaced by one bond whose conductivity is the same as the sum of the conductivities of the n bonds. We denote the conductivity of bond b_{ij} joining vertex x_i to vertex x_j by c_{ij}. As in section 2, we distinguish two vertices $x_1 = x$ and $x_M = y$, with $c_{1M} = 0$, and we are interested in the effective conductivity of G (with bond conductivities c_{ij}) measured between vertices x and y, denoted by $\sigma_{xy}(G)$. Our subsequent calculations are based on the following exact formula [**18**,**19**], which in some form was known to Kirchoff himself,

$$\sigma_{xy}(G) = \frac{\sum\limits_{T} \prod\limits_{b_{ij} \epsilon T} c_{ij}}{\sum\limits_{T_{xy}} \prod\limits_{b_{ij} \epsilon T_{xy}} c_{ij}} , \tag{4.1}$$

where the sum in the numerator is over all spanning trees in G, and the sum in the denominator is over all spanning trees T_{xy} in G with the vertices x and y identified. Another way of viewing the trees T_{xy} is by considering a new graph G' obtained from G by adding a bond b_{1M} between x and y (the battery bond) with $c_{1M} = h$. The set of trees T_{xy} are then those trees of G' which include b_{1M}, and the denominator of (4.1) is obtained by setting $h = 1$ for those trees.

To implement formula (4.1) on the computer, it is useful to write it in terms of an adjacency matrix [20]. Let the $M \times M$ matrix A be defined by

$$
\begin{aligned}
A_{ij} &= -c_{ij} \quad , \quad i,j = 1,\dots,M \\
A_{ii} &= \sum_{j \neq i} c_{ij} \quad , \quad i,j = 1,\dots,M \quad ,
\end{aligned}
\tag{4.2}
$$

with $c_{1M} = h$. We define two associated matrices as follows. Let A' be the $(M-1) \times (M-1)$ matrix formed from A by removing the first row and first column, corresponding to $x_1 = x$. Then let A'' be the $(M-2) \times (M-2)$ matrix formed from A' by subsequently removing its last row and last column, corresponding in A to $x_M = y$. Finally, define

$$
\begin{aligned}
Z(h) &= \det(A') \\
Z'(h) &= \det(A'')
\end{aligned}
\tag{4.3}
$$

where "det" means determinant. Then (4.1) can be written as

$$
\sigma_{xy}(G) = \left. \frac{\det(A')}{\det(A'')} \right|_{h=0} \tag{4.4}
$$

Recalling the statistical mechanics of the Ising model [see also ref. **21**], we can also write (4.4) as

$$
[\sigma_{xy}(G)]^{-1} = \frac{Z'(0)}{Z(0)} = \left(\frac{\partial}{\partial h} \ln Z \right)(0) \quad , \tag{4.5}
$$

which leads us to think of $Z(h)$ as a partition function, $\ln Z(h)$ as a free energy, and the resistance $[\sigma_{xy}(G)]^{-1}$ as a magnetization. Note that if all the bonds of G have unit conductivity then $Z(0)$ is the number of trees of G.

To calculate $\sigma_L^*(p)$, we let G be the graph described below (2.1), that is, G is an $L \times L$ sample of the $d = 2$ lattice, where L measures the number of vertices on a side. Included in G are the two vertices $x = x_1$ and $y = x_M$, which are attached to the vertices of two opposite sides by perfectly conducting bonds. Then $M = L^2 + 2$, and the number of bonds which have random conductivities 1 or ϵ is $N = 2\big((L-1)^2 + L - 1\big)$. To calculate the α_k in (2.1) we use the following procedure, which was carried out on a Sun 3/60 computer. First, α_0 is just the conductivity of G with the N bonds all assigned the conductivity 1. Then we randomly choose one bond and assign to it the conductivity $\epsilon = 0.001$ (ϵ must not be chosen too close to zero in order to avoid singular matrices), and calculate the resulting conductivity. Of the remaining 1's, we randomly choose another, and continue this procedure until all N bonds have conductivity ϵ. We repeat this procedure 30 times, letting $\overline{\alpha_k}$ be the average conductivity measured when there are k bonds of conductivity ϵ. Then we take

$$
\alpha_k = \binom{N}{k} \overline{\alpha_k} \ , \tag{4.6}
$$

which yields a reasonable numerical approximation to (2.1). The determinants in (4.4) are calculated by an IMSL routine. Plots of $\sigma_L^*(p)$ for complex p are obtained using Mathematica.

Using an analogous procedure to the above, we can define a polynomial $Z_L(p)$ for $h = 0$ by replacing the α_k above by

$$\beta_k = \binom{N}{k} \overline{\beta}_k \, , \tag{4.7}$$

where $\overline{\beta}_k$ is the average value of det $A'(k)/\det A'(0)$, $A'(k)$ is A' for G with k bonds of conductivity ϵ, and the normalization factor $1/\det A'(0)$ is to make $Z_L(1) = 1$ (otherwise $Z_L(1)$ is huge).

5. Results.

Our principal results are the figures below. Figure 1 (a-e) shows contour plots of $|\sigma_L^*(p)|$ for lattices with sizes ranging from 5×5 to 9×9. In each case, the range of $|\sigma_L^*(p)|$ shown is [0,2], and there are 10 contour levels. One of the main features of these plots is the persistence of a region W in which $|\sigma_L^*(p)|$ is very small. It is known, of course, that $\sigma^*(p) = 0$ for $0 \leq \mathrm{p} < p_c$. What is intriguing about the plot is that the vanishing appears to extend in the Im p direction to at least $|\mathrm{Im}\ \mathrm{p}\,| < 0.1$ in all cases shown, supporting the description of $\sigma^*(p)$ given in the Introduction. Figure 2 shows the fine structure in the 9×9 case, with a range for $|\sigma_9^*(p)|$ equal to [0, 0.1] with 10 contour levels again. Figure 3 shows a three dimensional plot of $|\sigma_L^*(p)|$ for the 9×9 case with vertical range [0,2] again, so that the surface describing $|\sigma_L^*(p)|$ has been chopped off above $|\sigma_L^*(p)| = 2$, giving the flat top in the outer fringes seen in the picture.

Here we should make a technical remark about "p_c," i.e., the threshold below which $\sigma_L^*(p)$ "vanishes" for real p. Since we are using $\epsilon = 0.001$ and finite L, this threshold is somewhat ill-defined. Furthermore, since we are averaging over realizations of sequential "removals," as long as there is one realization where a conducting path remains as k is increased beyond $N/2$ (or as p is decreased below $p = \frac{1}{2}$), this realization will dominate the average and produce an abnormally low "p_c," as seen in Figure 2. We chose to leave these realizations in to preserve randomness, and because we are only interested in the qualitative behavior of $\sigma^*(p)$, not in determining p_c, which is known to be $\frac{1}{2}$.

One feature of the graph of $|\sigma_L^*(p)|$ which can be seen in Figure 1 to a certain extent, and to a greater extent in plots having a much larger range, is that at the outer fringes of the plots, $|\sigma_L^*(p)|$ is growing very rapidly – the contours are very close together. At this point it is not clear if $\sigma^*(p)$ really has some sort of pole structure on this outer rectangular ring, as indicated in the 8×8 and 9×9 case, or if this is a manifestation of the series expansion (2.1) that we have used. It would be very interesting to plot $\sigma^*(p)$ via (3.2) with $s = 1$ to see if this divergent structure persists. However, the determination of the $\mu_n(p)$ is somewhat more involved than finding the α_k in (2.1).

In addition to the figures shown here, we have analyzed the zeros of $\sigma_L^*(p)$ by plotting the zero contours of Re$[\sigma_L^*(p)]$ and Im$[\sigma_L^*(p)]$, and finding where they intersect. The zeros seem to congregate on the fringes in the divergent structure, and somewhat in the region W. We have found, however, no evidence of the zeros lying on any particular curves through p_c, as would be the case in a Lee-Yang type transition.

Finally, in Figure 4 we show a contour plot of the fine structure (range = [0,0.1]) of $|\widetilde{Z}_L(p)|$ for $L = 9$, where $\widetilde{Z}_L(p)$ is defined by (2.1) with the α_k replaced by $(\beta_k)^{1/15}$. (We have taken the 15th root in particular because it forces the transition for $\widetilde{Z}_L(p)$ to occur approximately at $p = \frac{1}{2}$.) The region W for the partition function is similar to that for $\sigma_L^*(p)$, but appears to be somewhat larger.

Acknowledgements. I would like to thank G. Milton, P. Doyle, C. Newman, and C. Lobb for valuable conversations. I am also grateful to A. Hosoi for programming assistance, and to S. Barbu for preparing the manuscript.

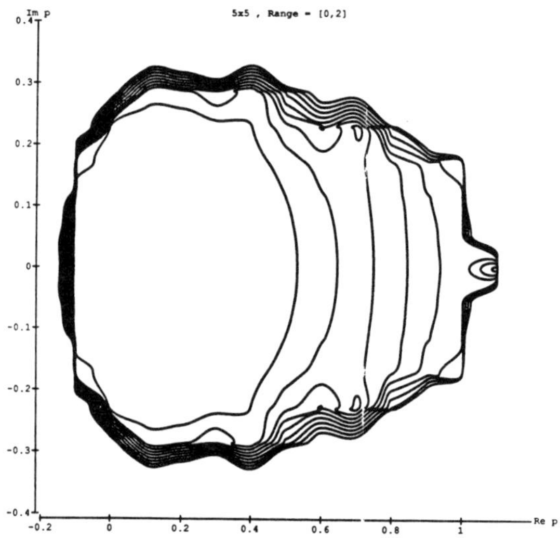

Figure 1a.

Figure 1(a-e). Contour plots of $|\sigma_L^*(p)|$ for $L = 5,\ldots,9$, with 10 contour levels over a range of [0,2] in each case.

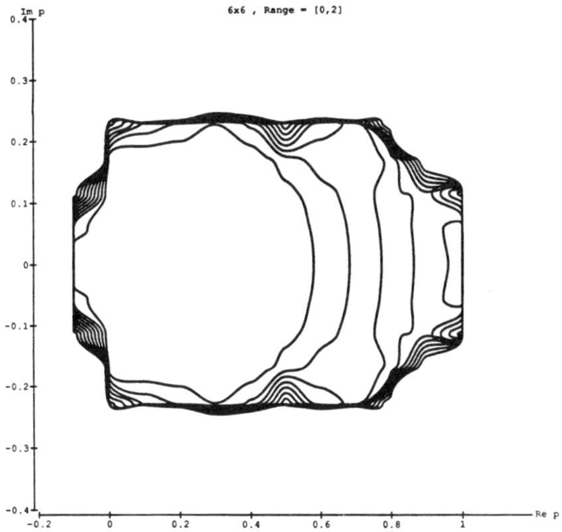

Figure 1b.

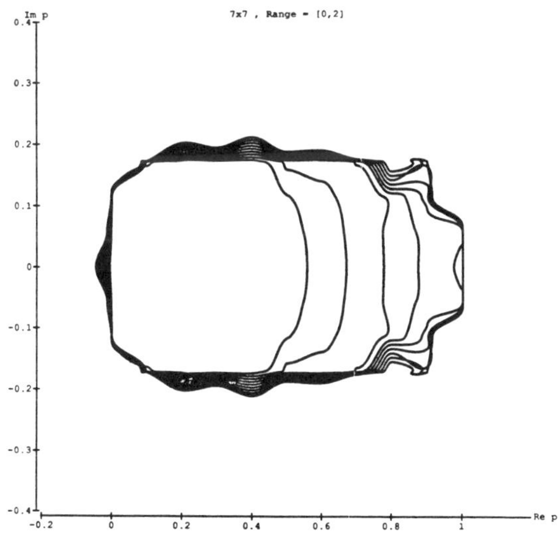

Figurc 1c.

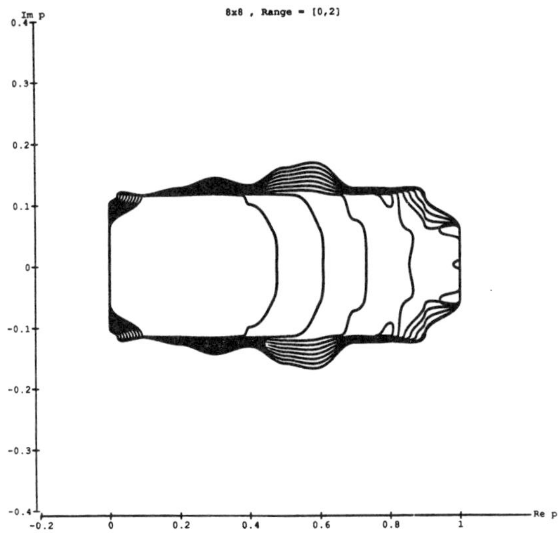

Figure 1d.

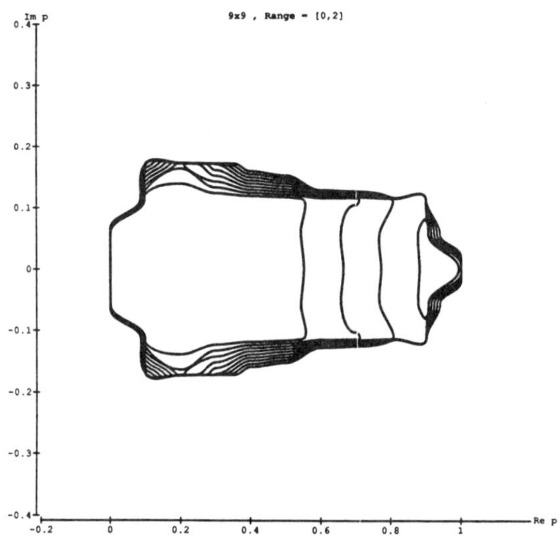

Figure 1e.

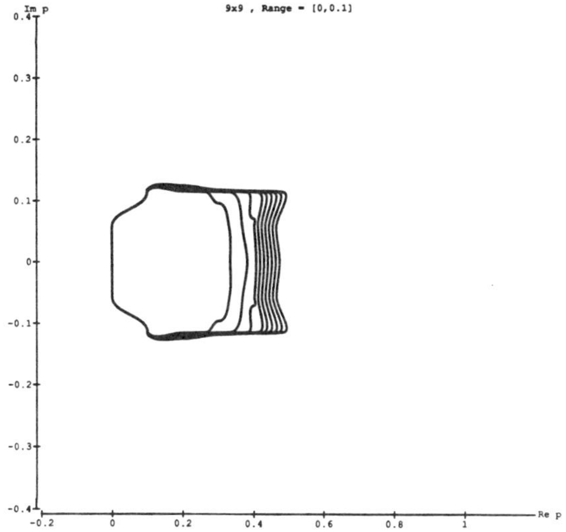

Figure 2. Contour plot of $|\sigma_L^*(p)|$ for $L = 9$ with 10 contour levels over a range of [0,0.1].

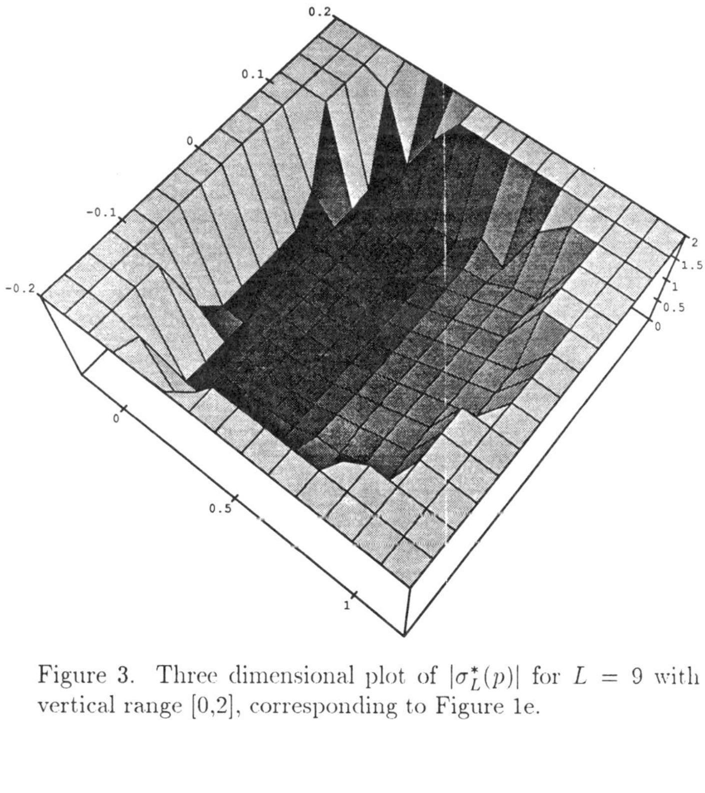

Figure 3. Three dimensional plot of $|\sigma_L^*(p)|$ for $L = 9$ with vertical range [0,2], corresponding to Figure 1e.

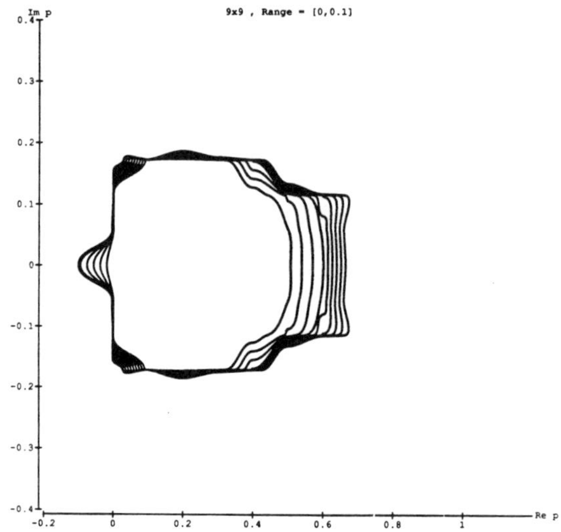

Figure 4. Contour plot of $|\widetilde{Z}_L(p)|$ for $L = 9$ with 10 contour levels over a range of [0,0.1].

REFERENCES

1. S. Kirkpatrick, Rev. Mod. Phys. **45**, 574 (1973).
2. D. Stauffer, Phys. Rep. **54**, 1 (1979).
3. H. Kesten, *Percolation Theory for Mathematicians*, Birkhaüser, Boston, 1982.
4. G. Grimmett and H. Kesten, Z. Wahr. **66**, 335 (1984).
5. J. T. Chayes and L. Chayes, Comm. Math. Phys. **105**, 133 (1986).
6. K. Golden, in *Random Media and Composite Materials*, R. V. Kohn and G. W. Milton, Eds., SIAM, Philadelphia, 1989.
7. C. J. Thompson, *Mathematical Statistical Mechanics*, Princeton Univ. Press, Princeton, 1972.
8. S. M. Kozlov, Dokl. Akad. Nauk. SSSR **241**, 1016 (1978).
9. G. Papanicolaou and S. Varadhan, in *Colloquia Mathematica Societatis János Bolyai* 27, *Random Fields*, North Holland, Amsterdam, 1982.
10. K. Golden and G. Papanicolaou, Comm. Math. Phys. **90**, 473 (1983).
11. R. Künnemann, *Effective conductivity on a lattice as a limit of box conductivities*, in Ph.D. Thesis, U. of Heidelberg (1983).
12. V. V. Zhikov, Math. Zametki **45**, 34 (1989).
13. D. J. Bergman, Phys. Rep. **C43**, 377 (1978).

14. O. Bruno, *The Effective Conductivity of an Infinitely Interchangeable Mixture*, Ph.D. Thesis, New York University (1989).
15. M. N. Miller, J. Math. Phys. **10**, 1988 (1969).
16. O. Bruno and K. Golden, J. Stat. Phys. **61**, 365 (1990).
17. A. S. Sangani and A. Acrivos, Proc. R. Soc. Lond. **A386**, 263 (1983).
18. S. J. Mason and H. J. Zimmermann, *Electronic Circuits, Signals, and Systems*, John Wiley and Sons, New York, 1960.
19. C. Thomassen, "Resistances and currents in infinite electrical networks," to appear in J. Comb. Th. B.
20. H. N. V. Temperley, *Graph Theory and Applications*, Ellis Horwood Ltd., West Sussex, 1981.
21. D. Coppersmith, P. Doyle, P. Raghavan, and M. Snir, in *Proc. of 22nd Ann. ACM Symp. on Theory of Computing*, Balt., Md., May 1990.

A STOCHASTIC FUBINI THEOREM AND EQUIVALENCE OF EXTENDED SOLUTIONS OF STOCHASTIC EVOLUTION EQUATIONS IN HILBERT SPACE

Luis G. Gorostiza and Jorge A. León
Centro de Investigación y de Estudios Avanzados
México 07000 D.F., México

1. INTRODUCTION

Let $X \equiv \{X_t,\ 0 \leq t \leq T\}$ be a process with values in a real separable Hilbert space H, which satisfies the stochastic evolution equation

$$\begin{aligned} dX_t &= A(t)X_t\,dt + dZ(X)_t, \quad 0 \leq t \leq T, \\ X_0 &= \xi, \end{aligned} \tag{1.1}$$

where $\{A(t),\ 0 \leq t \leq T\}$ is a family of closed linear operators on H which generates an evolution system $\{U(t,s),\ 0 \leq s \leq t \leq T\}$ on H, and $Z(X)$ is an H-valued semimartingale, 0 at time 0, which may depend on X (e.g. a stochastic integral of $X$with respect to another semimartingale). The variation of constants formula corresponding to (1.1) is formally written as

$$X_t = U(t,0)\xi + \int_0^t U(t,r)\,dZ(X)_r, \quad 0 \leq t \leq T, \tag{1.2}$$

which is also an equation for X (it is an explicit expression for the solution of (1.1) if Z does not depend on X).

Equations (1.1) and (1.2) are symbolic and they may be made precise in several ways. Assuming existence and uniqueness of solutions of (1.1) and (1.2) according to several possible definitions, the main question is how are these solutions related and, in particular, under what conditions are they equivalent. This problem was investigated by Chojnowska-Michalik [1,2] in the case where $A(t)$ does not depend on t and generates a semigroup on H, and Z is a semimartingale of a particular class. In this paper we are interested in weak forms of solutions of (1.1) and (1.2), and in equivalence results under weak requirements on the family $\{A(t)\}$ and for a general semimartingale Z.

Denoting by $\langle \cdot\,, \cdot \rangle$ the inner product in H, the adjoint integrated forms of (1.1) and (1.2) are

$$\langle X_t, y\rangle = \langle \xi, y\rangle + \int_0^t \langle X_r, A^*(r)y\rangle\,dr + \langle Z(X)_t, y\rangle \tag{1.3}$$

for

$$y \in \mathcal{D} \equiv \bigcap_{0 \leq t \leq T} \mathcal{D}(A^*(t)), \tag{1.4}$$

where $\mathcal{D}(A^*(t))$ denotes the domain of the adjoint $A^*(t)$ of $A(t)$, and

$$\langle X_t, y \rangle = \langle \xi, U^*(t,0)y \rangle + \int_0^t \langle U^*(t,r)y, dZ(X)_r \rangle, \quad y \in H. \tag{1.5}$$

A reason for considering equation (1.3) is that it does not need that X_t belong to $\mathcal{D}(A(t))$, the domain of $A(t)$, which is a requirement for strong solutions. A reason for considering equation (1.5) is that the stochastic integral in (1.2) may not be well-defined if $\{U(t,s)\}$ is not Borel-measurable, but the stochastic integral in (1.5) is well-defined if $\{U(t,s)\}$is only weakly-measurable. Equations (1.3) and (1.5) are natural weak forms of (1.1) and (1.2), and it is clear that they are not equivalent. Indeed, (1.3) requires that y belong to $\mathcal{D}$, defined by (1.4), whereas (1.5) puts no condition on the domains $\mathcal{D}(A^*(t))$. For (1.3) it is natural to require that $\mathcal{D}$ be dense in H, so that X_t is well-defined by the right-hand side of the equation. Under this condition and suitable additional assumptions, relationships and equivalence of solutions of (1.3) and (1.5) were established by León [7].

The question that interests us here is if it is still possible to obtain relationships between weak forms of solutions of (1.1) and (1.2) if $\mathcal{D}$ is no longer required to be dense in H. In order to address this question one has to extend the usual relationships between $\{A(t)\}$ and $\{U(t,s)\}$, as well as equation (1.3). This was done in the deterministic setting by Dawson and Gorostiza [3], and for stochastic equations by Gorostiza and León [4]. However, in [4] the main tools were extended forms of stochastic integration by parts formulas ([4], Lemmas 3.10 and 3.14), which imposed some uniformity requirements on the control processes of certain semimartingales, and some bounded variation conditions. Such restrictions are consistent with the usual hypotheses of semimartingale calculus [10], but they are undesirable. Thanks to a new stochastic Fubini theorem [8], it is possible to get rid of these conditions by means of a proof which does not use integration by parts formulas (the method of [1,2] also used a special stochastic Fubini theorem). In this paper we shall give the improved equivalence results and sketch the proofs based on the stochastic Fubini theorem. These proofs are significantly simpler than those of [4]; most of the stochastic calculus technicalities are contained in the proof of the stochastic Fubini theorem.

The processes below are defined on a stochastic basis $(\Omega, (\Im_t)_{0 \leq t \leq T}, \Im, P)$ satisfying the usual conditions.

2. STOCHASTIC FUBINI THEOREM

We will assume that μ is a non-negative σ-finite measure on the measurable space $(S, \mathcal{S})$, G is a real separable Hilbert space, $(L(H,G), \|\cdot\|)$ is the Banach space of bounded linear operators from H into G, $\mathcal{B}_{L(H,G)}$ is its Borel σ-algebra, and $\mathcal{P}$ is the predictable σ-algebra.

The following stochastic Fubini theorem gives a generalization to those of [1,2,5,6,9].

THEOREM. (León [8]). *Let Z be an H-valued semimartingale and $\varphi : S \times R^+ \times \Omega \to L(H,G)$ a Borel $\mathcal{S} \otimes \mathcal{P} - \mathcal{B}_{L(H,G)}$-measurable function, and assume*

(i) There exists a non-negative function $f \in L^1(\mu)$ such that $\|\varphi(s,r,\omega)\| \leq f(s)$ for all $(s,r,\omega) \in \mathcal{S} \times R^+ \times \Omega$.

(ii) For each $(r,\omega) \in R^+ \times \Omega$ the Bochner integral $\int_S \varphi(s,r,\omega)\,\mu(ds)$ is well-defined.

(iii) The process $\int_S \varphi(s,\cdot)\,\mu(ds)$ is Z-integrable.

Then for each $t \in R^+$ there exists a measurable function $Y^t : S \times \Omega \to G$, belonging to $L^1(\mu)$ w.p.1, such that

$$\int_{]0,t]} \varphi(s,r)\,dZ_r = Y_s^t \quad \textit{w.p.1 for almost all} \quad s \in S,$$

and

$$\int_{]0,t]} \int_S \varphi(s,r)\,\mu(ds)\,dZ_r = \int_S Y_s^t\,\mu(ds) \quad \textit{w.p.1.}$$

When $G = R$, the Pettis theorem, the Fubini theorem for Bochner integrals and hypothesis (i) imply that assumptions (ii) and (iii) hold, and if the R-valued function $\int_{]0,t]} \varphi(\cdot,r)\,dZ_r$ is $\mathcal{S} \otimes \Im$-measurable, then

$$Y_\cdot^t \equiv \int_{]0,t]} \varphi(\cdot,r)\,dZ_r.$$

3. EXTENDED EVOLUTION OPERATORS AND EXTENDED SOLUTIONS

We start by giving an approximation which is the main tool for the case where $\mathcal{D}$ is not necessarily dense in H. This is taken from [3].

Let $\{A(t),\ 0 \le t \le T\}$ be a family of closed linear operators on H such that the set

$$\mathcal{D}(\mathcal{A}) = \{f \in L^2([0,T],H) : f(t) \in \mathcal{D}(A(t)) \text{ for } 0 \le t \le T \text{ and } A(\cdot)f(\cdot) \in L^2([0,T],H)\}$$

is dense in $L^2([0,T],H)$. (This holds for example if $\bigcap_{0 \le t \le T} \mathcal{D}(A(t))$ is dense in H and $A(\cdot)y \in L^2([0,T],H)$ for $y \in \bigcap_{0 \le t \le T} \mathcal{D}(A(t))$). We then define $\mathcal{A}$ on $\mathcal{D}(\mathcal{A})$ by

$$\begin{aligned} \mathcal{A} : L^2([0,T],H) &\to L^2([0,T],H) \\ f &\mapsto [\mathcal{A}f](t) \equiv A(t)f(t), \quad 0 \le t \le T, \quad f \in \mathcal{D}(\mathcal{A}). \end{aligned}$$

Then $\mathcal{A}$ is a closed operator, and the adjoint operator

$$\begin{aligned} \mathcal{A}^* : L^2([0,T],H) &\to L^2([0,T],H) \\ f &\mapsto [\mathcal{A}^* f](\cdot) \end{aligned}$$

also has dense domain $\mathcal{D}(\mathcal{A}^*)$.

$\mathcal{A}^*$ is an extension of the family $\{A^*(t),\ 0 \le t \le T\}$ in the sense that if $h \in L^2([0,T],H)$ is such that $h(t) \in \mathcal{D}(A^*(t))$ for $0 \le t \le T$ and $A^*(\cdot)h(\cdot) \in L^2([0,T],H)$, then $h \in \mathcal{D}(\mathcal{A}^*)$ and $[\mathcal{A}^* h](t) = A^*(t)h(t)$ for $0 \le t \le T$.

Hence, given any $y \in H$ and $\epsilon > 0$, there exists $h_y^\epsilon \in \mathcal{D}(\mathcal{A}^*)$ such that

$$\int_0^T |h_y^\epsilon(t) - y|^2 \, dt < \epsilon,$$

where $|\cdot|$ denotes the norm on H. Note that this is an approximation of a constant function in $L^2([0,T],H)$ by an element of $\mathcal{D}(\mathcal{A}^*)$. Thanks to this approximation, in the usual relationships between $\{A^*(t)\}$ and $\{U^*(t,s)\}$, namely (see [7] for details)

$$\int_s^t \langle x, U^*(r,s)A^*(r)y\rangle \, dr = \langle x, U^*(t,s)y\rangle - \langle x, y\rangle$$

and

$$\int_s^t \langle x, A^*(r)U^*(t,r)y\rangle \, dr = \langle x, U^*(t,s)y\rangle - \langle x, y\rangle,$$

we can replace $A^*(r)y$ and $A^*(r)U^*(t,r)y$ by elements of $\mathcal{D}(\mathcal{A}^*)$ and write the errors thus made in additive form.

DEFINITION 3.1. $\{U^*(t,s)\}$ is an *extended weak forward adjoint evolution operator* (EWFA) if for every $x, y \in H$ and $\epsilon > 0$ there exists $h_y^\epsilon \in \mathcal{D}(\mathcal{A}^*)$ such that

$$\int_s^t \langle x, U^*(r,s)[\mathcal{A}^* h_y^\epsilon](r)\rangle\, dr = \langle x, U^*(t,s)y\rangle - \langle x, y\rangle + e_{x,y}^{1,\epsilon}(s,t)$$

for all $0 \le s < t \le T$, where $e_{x,y}^{1,\epsilon}(s,t) \to 0$ as $\epsilon \to 0$ uniformly for $s \in [0,t]$ and x in bounded sets in H.

DEFINITION 3.2. $\{U^*(t,s)\}$ is an *extended weak backward adjoint evolution operator* (EWBA) if for every $x, y \in H$ and $\epsilon > 0$ there exists $h_{U^*(t,r)y}^\epsilon \in \mathcal{D}(\mathcal{A}^*)$, for each $r \in [s,t]$, satisfaying $(r \mapsto [\mathcal{A}^* h_{U^*(t,r)y}^\epsilon](r)) \in L^2([s,t],H)$, and

$$\int_s^t \langle x, [\mathcal{A}^* h_{U^*(t,r)y}^\epsilon](r)\rangle\, dr = \langle x, U^*(t,s)y\rangle - \langle x, y\rangle + e_{x,y}^{2,\epsilon}(s,t)$$

for all $0 \le s < t \le T$, where $e_{x,y}^{2,\epsilon}(s,t) \to 0$ as $\epsilon \to 0$ uniformly for $s \in [0,t]$ and x in bounded sets in H.

For our results we will need the following assumption on EWBA, which may seem unnatural but we shall see that it is not.

ASSUMPTION 3.3. Let $\{U^*(t,s)\}$ be an EWBA. We assume that there exists a dense subset D of H such that given $y \in D$, $\epsilon > 0$ and $h_{U^*(t,r)y}^\epsilon \in \mathcal{D}(\mathcal{A}^*)$ satisfying Definition 3.2 for each $r \in [s,t]$, there exists $h^\epsilon_{[\mathcal{A}^* h^\epsilon_{U^*(t,r)y}](r)} \in \mathcal{D}(\mathcal{A}^*)$ such that

$$((r,r') \mapsto [\mathcal{A}^* h^\epsilon_{[\mathcal{A}^* h^\epsilon_{U^*(t,r)y}](r)}](r')) \in L^1([s,t]\times[s,t],H)$$

and

$$\int_s^t [\mathcal{A}^* h^\epsilon_{[\mathcal{A}^* h^\epsilon_{U^*(t,r)y}](r)}](s)\, dr = [\mathcal{A}^* h^\epsilon_{U^*(t,s)y}](s) - [\mathcal{A}^* h_y^\epsilon](s) + e_y^{3,\epsilon}(s,t)$$

for all $s < t$, where $e_y^{3,\epsilon}(s,t) \to 0$ as $\epsilon \to 0$ and is uniformly bounded for $s \in [0,t]$.

Similarly as above, in equation (1.3) we can replace $A^*(r)y$ by an approximation in $\mathcal{D}(\mathcal{A}^*)$ and write the error thus made in additive form. $\mathcal{N}_0$ below is a subclass of the space of H-valued progressively measurable processes on which the semimartingale space-valued mapping Z is well-defined.

DEFINITION 3.4. An H-valued process X is an *extended *-solution* of equation (1.1) if and only if

(*i*) $X \in \mathcal{N}_0$.
(*ii*) $X_\cdot \in L^1([0,T],H)$ w.p.1.
(*iii*) For each $y \in H$ and $\epsilon > 0$ there exists $h_y^\epsilon \in \mathcal{D}(\mathcal{A}^*)$ such that $\langle X_\cdot, [\mathcal{A}^* h_y^\epsilon](\cdot)\rangle \in L^1([0,T],R)$ w.p.1.
(*iv*) For each $t \in [0,T]$, $y \in H$ and $\epsilon > 0$,

$$\langle X_t, y\rangle = \langle \xi, y\rangle + \int_0^t \langle X_r, [\mathcal{A}^* h_y^\epsilon](r)\rangle\, dr + \langle Z(X)_t, y\rangle + e_y^\epsilon(0,t) \quad \text{w.p.1},$$

where $e_y^\epsilon(0,t) \to 0$ in probability as $\epsilon \to 0$.

We remark that when $\mathcal{D}$ is dense in H, the definitions and assumption above reduce to the restricted ones, without error terms [7].

Finally we define (without extension!) the version we need of equation (1.5).

DEFINITION 3.5. An H-valued process X is a **-evolution solution* of equation (1.1) if and only if

(*i*) $X \in \mathcal{N}_0$.
(*ii*) $X_\cdot \in L^1([0,T],H)$ w.p.1.
(*iii*) For each $t \in [0,T]$ and $y \in H$,

$$\langle X_t, y\rangle = \langle \xi, U^*(t,0)y\rangle + \int_{]0,t]} \langle U^*(t,r)y, dZ(X)_r\rangle \quad \text{w.p.1.}$$

REMARK. In [4] we defined an extended *-evolution solution, which is an extended version of the last definition, involving an approximation of $U^*(t,r)y$ of the form $h^\epsilon(r)U^*(t,r)y$, where $h^\epsilon(\cdot)$ is $L(H,H)$-valued, so as to satisfy a bounded variation requirement, plus an error term. This is now unnecessary.

The following results from [4] will be needed.

LEMMA 3.6. *Let* $\{U^*(t,s)\}$ *be an EWFA. Then for each* $y \in H$, $t \in [0,T]$, $\epsilon > 0$ *and* $h_y^\epsilon \in \mathcal{D}(\mathcal{A}^*)$ *satisfying Definition 3.1, there exists a measurable function* $e_y^{1,\epsilon}(\cdot,t)$:

$[0,t] \to H$ *such that*

$$\int_s^t U^*(r,s)[\mathcal{A}^* h_y^\epsilon](r)\,dr = U^*(t,s)y - y + e_y^{1,\epsilon}(s,t),$$

for $s \in [0,t]$, *and* $e_y^{1,\epsilon}(s,t) \to 0$ *as* $\epsilon \to 0$ *uniformly for* $s \in [0,t]$.

LEMMA 3.7. *Let* $\{U^*(t,s)\}$ *be an EWBA. Then for each* $y \in H$, $t \in [0,T]$, $\epsilon > 0$ *and* $h^\epsilon_{U^*(t,\cdot)y}$ *satisfying Definition 3.2, there exists a measurable function* $e_y^{2,\epsilon}(\cdot,t) : [0,t] \to H$ *such that*

$$\int_s^t [\mathcal{A}^* h^\epsilon_{U^*(t,r)y}](r)\,dr = U^*(t,s)y - y + e_y^{2,\epsilon}(s,t), \tag{3.1}$$

for $s \in [0,t]$, *and* $e_y^{2,\epsilon}(s,t) \to 0$ *as* $\epsilon \to 0$ *uniformly for* $s \in [0,t]$.

LEMMA 3.8. *Let* $\{U^*(t,s)\}$ *be an EWBA and* Z *an* H*-valued semimartingale*, $Z_0 = 0$. *Then for each* $y \in H$, $t \in [0,T]$ *and* $\epsilon > 0$,

$$\begin{aligned}\int_0^t \langle Z_r, [\mathcal{A}^* h^\epsilon_{U^*(t,r)y}](r)\rangle\,dr &= \int_{]0,t]} \langle U^*(t,r)y, dZ_r\rangle \\ &+ \int_{]0,t]} \langle e_y^{2,\epsilon}(r,t), dZ_r\rangle - \langle Z_t, y\rangle \quad \text{w.p.1},\end{aligned} \tag{3.2}$$

where $h^\epsilon_{U^*(t,r)y}$ *and* $e_y^{2,\epsilon}(r,t)$ *satisfy* (3.1).

The last lemma is a consequence of the integration by parts formula [4] (Lemma 3.10), but now it can be proved under weaker assumptions with the help of the stochastic Fubini theorem.

PROOF. Fix $y \in H$, $t \in [0,T]$, $\epsilon > 0$ and $h^\epsilon_{U^*(t,\cdot)y}$ satisfying Definition 3.2. Let $r \in [0,t]$. Then

$$\langle Z_r, [\mathcal{A}^* h^\epsilon_{U^*(t,r)y}](r)\rangle = \int_{]0,r]} \langle [\mathcal{A}^* h^\epsilon_{U^*(t,r)y}](r), dZ_{r'}\rangle.$$

Hence the stochastic Fubini theorem implies

$$\int_0^t \langle Z_r, [\mathcal{A}^* h^\epsilon_{U^*(t,r)y}](r)\rangle\,dr$$

$$= \int_{]0,t]} \int_{r'}^{t} \langle [\mathcal{A}^* h^{\epsilon}_{U^*(t,r)y}](r), \cdot \rangle \, dr \, dZ_{r'} \quad \text{w.p.1.}$$

Finally, by Lemma 3.7 there exists a measurable function $e_y^{2,\epsilon}(\cdot, t) : [0,t] \to H$ satisfying (3.1) and (3.2).

We can now state our results.

4. EQUIVALENCE THEOREMS

THEOREM 4.1. *Let X be an extended *-solution and $\{U^*(t,s)\}$ an EWBA satisfying Assumption 3.3 such that for every $t \in [0,T]$, $y \in D$ and $\epsilon > 0$ the process*

$$r \mapsto e^{\epsilon}_{[\mathcal{A}^* h^{\epsilon}_{U^*(t,r)y}](r)}(0, r)$$

has a measurable version $e_y^{\epsilon, t}(\cdot)$ on $[0,t]$ and $[\mathcal{A}^* h^{\epsilon}_{[\mathcal{A}^* h^{\epsilon}_{U^*(t,r)y}](r)}](r')$ remains bounded for $(r, r') \in [0,t] \times [0,t]$. Assume in addition that for every $t \in [0,T]$ and $y \in D$, $\int_0^t e_y^{\epsilon, t}(r) \, dr \to 0$ in probability as $\epsilon \to 0$.

*Then X is a *-evolution solution.*

This theorem is proved using Lemmas 3.7 and 3.8 the same way as in [4] Theorem 3.5 was proved by the corresponding results: Proposition 2.4 and Lemma 3.10. The only difference is that due to the stochastic Fubini theorem, Lemma 3.8 holds under weaker assumptions than Lemma 3.10 of [4].

THEOREM 4.2. *Let X be a *-evolution solution and $\{U^*(t,s)\}$ an EWFA. Then X is an extended *-solution.*

This theorem improves Theorem 3.6 of [4], which requires an extended version of the *-evolution solution and a bounded variation condition in order to apply the integration by parts formula.

PROOF. The fact that X is a *-evolution solution and the stochastic Fubini theorem imply

$$\int_0^t \langle X_r, [\mathcal{A}^* h^{\epsilon}_y](r) \rangle \, dr = \int_0^t \langle \xi, U^*(r,0)[\mathcal{A}^* h^{\epsilon}_y](r) \rangle \, dr$$

$$+ \int_{]0,t]} \int_{r'}^{t} \langle U^*(r,r')[\mathcal{A}^* h_y^\epsilon](r), \cdot\rangle \, dr \, dZ(X)_{r'} \quad \text{w.p.1.}$$

Then, by Lemma 3.6 we obtain

$$\int_0^t \langle X_r, [\mathcal{A}^* h_y^\epsilon](r)\rangle \, dr = \langle \xi, U^*(t,0)y - y\rangle$$

$$-\langle Z(X)_t, y\rangle + \int_{]0,t]} \langle U^*(t,r)y, dZ(X)_r\rangle$$

$$+\langle \xi, e_y^{1,\epsilon}(0,t)\rangle + \int_{]0,t]} \langle e_y^{1,\epsilon}(r,t), dZ(X)_r\rangle \quad \text{w.p.1.}$$

Finally, since X is a *-evolution solution, the fact that $e_y^{1,\epsilon}(r,t) \to 0$ uniformly for $r \in [0,t]$ and the stochastic dominated convergence theorem imply that X is also an extended *-solution.

The following consequences are immediate.

COROLLARY 4.3. *Under the assumptions of Theorems 4.1 and 4.2, extended *-solutions and *-evolution solutions are equivalent.*

COROLLARY 4.4. *In the linear case (i.e., when Z does not depend on X), the assumptions of Corollary 4.3 imply that if $U(t,\cdot)$ is Z-integrable on $[0,t]$ for all $t \in [0,T]$, then there is a unique extended *-solution of equation (1.1) and it is given by*

$$X_t = U(t,0)\xi + \int_0^t U(t,r)\, dZ_r \quad \textit{for each} \quad t \in [0,T].$$

REMARKS.

1. As shown in [3], even for simple examples of the form

$$A(t) = I_{[0,T/2]}(t)A_1 + I_{]T/2,T]}(t)A_2,$$

where A_1 and A_2 generate self-adjoint contraction semigroups $\{T_t^{(1)}\}$ and $\{T_t^{(2)}\}$ on H, the extended formulation includes cases not covered by the restricted formulation [7]. This is because A_1 and A_2 can be chosen so that for $t > 0$, $T_t^{(2)}$ does not map $\mathcal{D}(A_2)$ into $\mathcal{D}(A_1)$.

2. Assumption 3.3 reduces to the corresponding one in the restricted case [3,7], and it is automatically satisfied when $A(t)$ does not depend on t. In [3] there is an example of an equation where Assumption 3.3 does not hold (because the error term does not go to 0), and there are two different extended *-solutions and a single *-evolution solution, which is a counterexample to Theorem 4.1 without this assumption.

References

[1] Chojnowska-Michalik, A., Stochastic differential equations in Hilbert space and some of their applications. Ph.D. thesis, Institute of Mathematics, Polish Academy of Sciences, 1976.

[2] Chojnowska-Michalik, A., Stochastic differential equations in Hilbert space. *Probability Theory-Banach Center Publications*, *5*, (1979), 53-74.

[3] Dawson, D.A. and Gorostiza, L.G., *-solutions of evolution equations in Hilbert space. *Journal of Differential Equations*, *68*, (1987), 299-319.

[4] Gorostiza, L.G. and León, J.A., *-solutions of stochastic evolution equations in Hilbert space. *Rep. Int. 29*, Depto. Matemáticas. CINVESTAV (1989).

[5] Ikeda, N. and Watanabe, S., Stochastic Differential Equations and Diffusion Processes. North-Holland, Amsterdam-Oxford-New York, 1981.

[6] Jacod, J., Calcul Stochastique et Problèmes de Martingales. *Lec. Notes in Math. 714*, Springer-Verlag, Berlin-New York-Heidelberg, 1979.

[7] León, J.A., Stochastic evolution equations with respect to semimartingales in Hilbert space. *Stochastics, 27*, (1989), 1-21.

[8] León, J.A., Stochastic Fubini theorem for semimartingales in Hilbert space. *Canadian Journal of Mathematics* (to appear).

[9] Liptser, R.S. and Shiryayev, A.N., Statistics of Random Processes 1. General Theory, Springer-Verlag, New York-Heidelberg-Berlin, 1977.

[10] Metivier, M., Semimartingales. W. de Gruyter, Berlin-New York, 1982.

Hellinger Integrals and Hellinger Processes For Solutions of Martingale Problems

B. Grigelionis
Institute of Mathematics and Cybernetics
232600 Vilnius, Akademijos 4, Lithuania

1. Let $(\Omega, \mathcal{F})$ be a measurable space endowed with two probability measures P and P'. The Hellinger integral $H(P,P')$ of P and P' is the nonnegative number

$$H(P,P') = E_Q\left(\sqrt{\frac{dP}{dQ}\frac{dP'}{dQ}}\right),$$

where Q is any probability measure, such that $P \ll Q$ and $P' \ll Q$, and E_Q denotes the expectation with respect to Q. Starting with the famous S. Kakutani's paper [1], the role of the Hellinger integrals and the Kakutani-Hellinger distance $\rho(P,P')$, defined by means of the formula

$$\rho^2(P,P') = 1 - H(P,P'),$$

is increasing in the theory of statistical inference. Recall that the Kakutani-Hellinger distance defines the same topology as the variation metric $\|P - P'\|$ on the space of all probability measures on $(\Omega, \mathcal{F})$, because we have (see, e.g. [2]-[3]) that

$$2\rho^2(P,P') \leq \|P - P'\| \leq 2\sqrt{2}\rho(P,P').$$

This distance is naturally used to define the regularity properties of the statistical experiments and the asymptotical behaviour of the maximal likelihood estimators of the unknown parameters or the error probabilities of the optimal statistical tests (see, e.g., [3]-[6]).

Introducing the Hellinger integral of complex valued order z by

$$H(z;P,P') = E_Q\left[\left(\frac{dP}{dQ}\right)^z\left(\frac{dP'}{dQ}\right)^{1-z}\right], 0 \leq Rez \leq 1,$$

under assumption that $P \sim P'$ we easily find a formula for the characteristic function of the log-likelihood ratio in terms of the related Hellinger integrals (see [3]):

$$E_{p'} \exp\{ivln\frac{dP}{dP'}\} = \lim_{u\downarrow 0} H(u+iv, P, P').$$

Finally, we remark, that the important properties of contiguity and entire separation of the sequence $\{P'^n, n \geq 1\}$ to the sequence $\{P^n, n \geq\}$ of the probability measures on the measurable spaces $(\Omega^n, \mathcal{F}^n)$, respectively, can be expressed in terms of the Hellinger integrals of order $\alpha, 0 < \alpha < 1$. Namely (see, e.g.,[2], [7]), $\{P'^n, \geq 1\} \lhd \{P^n, n \geq 1\}$ iff

$$\lim_{\alpha\downarrow 0} \underline{\lim}_{n\to\infty} H(\alpha; P^N, P'^n) = 1,$$

and $\{P'^n, n \geq 1\}\Delta\{P^n, n \geq 1\}$ iff for some $\alpha \in (0,1)$

$$\underline{\lim}_{n\to\infty} H(\alpha; P^n, P'^n) = 0.$$

In the case, when $(\Omega^n, \mathcal{F}^n, P^n, P'^n) = (\Omega, \mathcal{F}, P, P')$ for all $n \geq 1$, we have the criteria for absolute continuity and singularity of measures P and P' in terms of the Hellinger integrals.

In this paper we shall consider the problem of evaluation of the Hellinger integrals for solutions of the defined martingale problems on filtered probability spaces. The notions of the Hellinger measures and the Hellinger processes, initiated by C.M. Newman [8], will be exploited.

2. Let $(\Omega, \mathcal{F}, F)$ be a filtered measurable space with a family $\{P^\theta, \theta \in \Theta\}$ of probability measures on $\mathcal{F}$.

For the sake of simplicity assume that there exists a probability measure Q on $\mathcal{F}$ such that for each $t > 0, \theta \in \Theta$

$$P_t^\theta = P^\theta|\mathcal{F}_t \sim Q_t = Q|\mathcal{F}_t \tag{1}$$

and $P_0^\theta = Q_0$.

Denote

$$Z_t^\theta = \frac{dP_t^\theta}{dQ_t}, t \geq 0, \theta \in \Theta,$$

$$\alpha = (\alpha_1, \ldots, \alpha_k), 0 < \alpha_j < 1, \sum_{j=1}^k \alpha_j = 1, k \geq 2,$$

$$I = (\theta_1, \ldots, \theta_k), \theta_j \in \Theta, j = 1, \ldots, k$$

$$Y_t(\alpha; I) = \prod_{j=1}^{k} (Z_t^{\theta_j})^{\alpha_j}, t \geq 0.$$

Because for each $\theta \in \Theta Z^\theta$ is a strictly positive (Q, F)-martingale, then according to the Hölder inequality for each $0 \leq s < t$ (cf. [9]).

$$E_Q[\frac{Y_t(\alpha; I)}{Y_s(\alpha; I)} | \mathcal{F}_s] = E_Q[\prod_{j=1}^{k} (\frac{Z_t^{\theta_j}}{Z_s^{\theta_j}})^{\alpha_j} | \mathcal{F}_s] \leq$$

$$\leq \prod_{j=1}^{k} [E(\frac{Z_t^{\theta_j}}{Z_s^{\theta_j}} | \mathcal{F}_s)]^{\alpha_j} = 1.$$

Thus $Y(\alpha, I)$ is a strictly positive $(Q, \mathbf{F})$-supermartingale. Using the Itô-Watanabe multiplicative decomposition theorem (see, e.g., [10]) we find that there exist a (unique up to Q-indistinguishability) decreasing $\mathbf{F}$-predictable process $D(\alpha; I), D_o(\alpha; I) = 1$, and a $(Q, \mathbf{F})$-local martingale $L(\alpha; I), L_o(\alpha; I) = 1$, such that

$$Y_t(\alpha; I) = D_t(\alpha; I) L_t(\alpha; I), t \geq 0. \tag{2}$$

Denote

$$h_t(\alpha; I) = -\int_0^t \frac{dD_s(\alpha; I)}{D_{s-}(\alpha; I)}, t \geq 0. \tag{3}$$

Definition 1 (cf. [9]). We call $h(\alpha; I)$ the Hellinger process of order α for the subset $\{P^\theta, \theta \in I\}$.

Definition 2 The measures

$$H_t(\alpha; I)(A) = E_Q[Y_t(\alpha; I) 1_A], A \in \mathcal{F}_t, t \geq 0, \tag{4}$$

we call the Hellinger measures of order α for the subset $\{P^\theta, \theta \in I\}$.

Assuming that $L(\alpha; I)$ is a $(Q, \mathbf{F})$-martingale, i.e.

$$E_Q L_t(\alpha; I) \equiv 1, \tag{5}$$

we define the probability measures

$$P_t(\alpha; I)(A) = E_Q[L_t(\alpha; I)1_A], A \in \mathcal{F}_t, t \geq 0. \tag{6}$$

From (2)-(6) we obtain that for each $t \geq 0$

$$H_t(\alpha; I) \sim P_t(\alpha; I)$$

and

$$\frac{dH_t(\alpha; I)}{dP_t(\alpha; I)} = \mathcal{E}_t(-h(\alpha; I)), \tag{7}$$

where $\mathcal{E}_t(\cdot)$ denotes the Doléans-Dade exponential.

In the case when the probability measures $P^\theta, \theta \in \Theta$, and Q are unique solutions to the defined martingale problems we shall identify the measures $P_t(\alpha; I)$ and the Hellinger processes $h_t(\alpha; I)$. Martingale representation theorems and properties of the Doléans-Dade exponentials will be applied.

3. Assume that $\mathbf{F} = \mathbf{F}^X$, where $\mathbf{F}^X$ denotes a filtration, generated by some m-dimensional cádlág process X, the measures $P^\theta, \theta \in \Theta$, and Q are uniquely characterized by their restrictions $P^\theta_o, \theta \in \Theta, Q_o$ and the properties, that X is a $(P^\theta, \mathbf{F})$-semimartingale and a $(Q, \mathbf{F})$-semimartingale with the triplets $(\delta^\theta, B, \sqcap^\theta)$ and $(\delta, B, \sqcap)$ of predictable characteristics[1], respectively, such that

$$\delta^\theta_t = \delta_t + \int_o^t g^\theta(s) B'(s) d\beta_s + \int_0^t \int_{0|x| \leq 1} x(\Psi^\theta(s,x) - 1) \sqcap (ds, dx),$$

$$\sqcap^\theta([0,t] \times \Gamma) = \int_0^t \int_\Gamma \Psi^\theta(s,x) \sqcap (ds, dx), \theta \in \Theta,$$

$$B_t = \|\beta^{jk}_t\|^m_{j,k=1}, \beta_t = \sum_{j=1}^m \beta^{jj}_t,$$

$$\beta^{jk'}_t = \frac{d\beta^{jk}_t}{d\beta_t}, B'_t = \|\beta^{jk'}_t\|^m_{j,k=1}, t \geq 0,$$

[1]for terminology and definitions see, e.g., [2], [10].

where Θ is an arbitrary set.

Let $p([0,t]\times\Gamma)=\sum_{s\le t}1_\Gamma(\Delta X_s)$ be the jump measure of $X, q([0,t]\times\Gamma)=p([0,t]\times\Gamma)-\sqcap([0,t]\times\Gamma), \Gamma\in B(E_m), E_m=R^m\backslash\{0\}$ and X^c be the continuous local martingale part of the $(Q,\mathbf{F})$-semimartingale X.

Denote $a(t)=\sqcap(\{t\}\times E_m), a^\theta(t)=\sqcap^\theta(\{t\}\times E_m)$,

$$a_I(t)=(a^{\theta_1}(t),\dots,a^{\theta_k}(t)),$$

$$a_{\alpha,I}(t)=\prod_{j=1}^k\widehat{(\Psi^{\theta_j})^{\alpha_j}}(t)$$

$$\left(\hat U(t)=\int_{E_m}U(tx)\sqcap(\{t\}\times dx)\right)$$

$$W^\theta_{\alpha_j}(t,x)=(\Psi^\theta(t,x))^{\alpha_j}-1+\frac{(\Psi^\theta)^{\alpha_j}(t)-a(t)}{1-a(t)}1_{\{a(t)<1\}},$$

$$W_{\alpha,I}(t,x)=\prod_{j=1}^k(\Psi^{\theta_j}(t,x))^{\alpha_j}-1+\frac{a_{\alpha,I}(t)-a(t)}{1-a(t)}1_{\{a(t)<1\}}.$$

$$g_{\alpha,I}(t)=\sum_{j=1}^k\alpha_j g^{\theta_j}(t),$$

$$\phi_\alpha(y)=\sum_{j=1}^k\alpha_j y_j-\prod_{j=1}^k(y_j)^{\alpha_j},$$

$$y=(y_1,\dots,y_k).$$

Assume that for each $t>0$ and $\theta\in\Theta$ $P^\theta_t\sim Q_t$,

$$W^{\theta_j}_{\alpha_j}\in g_{loc}(\sqcap,\mathbf{F}), j=1,\dots,k, W_{\alpha,I}\in g_{loc}(\sqcap,\mathbf{F}),$$

$$\{a = 1\} \subseteq \{a_{\alpha,I} = 1, a^{\theta j} = 1, \widehat{(\Psi^{\theta j})}^{\alpha j} = 1, j = 1, \ldots, k\}.$$

Theorem 1 *(cf. [2], [9]). Under above assumptions*

$$L_t(\alpha; I) = \mathcal{E}_t(N(\alpha; I)), t \geq 0,$$

where

$$N_t(\alpha; I) = \int_0^t g_{\alpha,I}(s) dX_s^c + \int_0^t \int_{E_m} W_{\alpha,I}(s,x) q(ds, dx),$$

and

$$h_t(\alpha; I) = \frac{1}{2}(\sum_{j=1}^k \alpha_j \int_0^t (g^{\theta j}(s), B'(s) g^{\theta j}(s)) d\beta_s -$$

$$- \int_0^t (g_{\alpha,I})(s), B'(s) g_{\alpha,I}(s)) d\beta_s +$$

$$+ \int_0^t \int_{E_m} \varphi_\alpha(\Psi_I(s,x)) \sqcap (ds, dx) + \sum_{s \leq t} \varphi_\alpha(1_k - a_I(s));$$

$$1_k = \underbrace{(1, \ldots, 1)}.$$

The proof of this theorem is based on the explicit formulas for the local densities Z_t^θ (see [2]), the properties of the stochastic integrals and the following lemma.

Let $(\mathcal{F}, B)$ be a Blackwell space, $(\Omega, \mathcal{F}, \mathbf{F}, P)$ be a stochastic basis, μ be an integer-valued random measure on $(R_+ \times X, B(R_+) \otimes B)$ with the $(P, \mathbf{F})$-compensator $\nu, \kappa(t) = \nu(< t > \times X)$,

$$\hat{V}(t) = \int_X V(t,x) \nu(\{t\} \times dx),$$

$$W(t,x) = V(t,x) - 1 + \frac{\hat{V}(t) - \kappa(t)}{1 - \kappa(t)} 1_{\{\kappa(t) < 1\}}, V(t,x) \geq 0,$$

$$W_\gamma(t,x) = V^\gamma(t,x) - 1 + \frac{\hat{V}^\gamma(t) - \kappa(t)}{1-\kappa(t)} 1_{\{\kappa(t)<1\}}.$$

Assuming that $W \in g_{loc}(\nu, \mathbf{F}), W_\gamma \in g_{loc}(\nu, \mathbf{F}), 0 < \gamma < 1, Y^c \in M^c_{loc}(P, \mathbf{F})$, denote

$$Y_t = Y^c_t + \int_0^t \int_X W(s,x)(\mu - \nu)(ds,dx),$$

$$Y^{(\gamma)}_t = \gamma Y^c_t + \int_0^t \int_X W_\gamma(s,x)(\mu - \nu)(ds,dx),$$

$$h_t(\gamma) = \frac{1}{2}\gamma(1-\gamma)\langle Y^c\rangle_t + \int_0^t \int_X \phi^{(\gamma)}(V(s,x),1)\nu(ds,dx) +$$

$$+ \sum_{s\leq t} \varphi^{(\gamma)}(1-\hat{V}(s), 1-\kappa(s)),$$

where

$$\varphi^{(\gamma)}(u,v) = \gamma u + (1-\gamma)v - u^\gamma v^{1-\gamma}.$$

Lemma 1 *If $\{\kappa = 1\} \subseteq \{\hat{V} = 1, \hat{V}^\gamma = 1\}$, then the following decomposition holds:*

$$\mathcal{E}^\gamma_t(Y) = \mathcal{E}_t(Y^{(\gamma)})\mathcal{E}_t(-h(\gamma)), t \geq 0.$$

Proof. Observe that under assumptions of the lemma we have that

$$(1 + \Delta Y_t)^\gamma = 1 + \Delta Y^{(\gamma)}_t, t > 0.$$

The proof completes after the standard calculations.

Theorem 2 *If additionally to the assumptions of theorem 1*

$$E_Q \mathcal{E}_t(N(\alpha; I)) \equiv 1,$$

then $P_t(\alpha; I) = P(\alpha; I)|\mathcal{F}_t, t \geq 0$, *and* $P(\alpha; I)$ *is uniquely characterized by the properties that* $P_0(\alpha; I) = Q_0$ *and* X *is a* $(P(\alpha; I), \mathbf{F})$*-semimartingale with the triplet* $(\delta(\alpha; I), B, \sqcap(\alpha; I))$ *of predictable characteristics, where*

$$\delta_t(\alpha; I) = \delta_t + \int_0^t g_{\alpha,I}(s)B'(s)d\beta_s +$$

$$+ \int_0^t \int_{|x|\leq 1} \times (\prod_{j=1}^k (\Psi^{\theta j}(s,x))^{\alpha j} - 1) \sqcap (ds, dx),$$

$$\sqcap(\alpha; I)([0,t] \times \Gamma) = \int_0^t \int_\Gamma \prod_{j=1}^k (\Psi^{\theta j}(s,x))^{\alpha_j} \sqcap (ds, dx).$$

The proof of this statement follows directly from the theorem 1 and the Girsanov's theorem for semimartingales.

4. As illustrations now we shall consider some examples.

Example 1 (exponential class of processes with the independent increments). Assume that

$$\frac{dP_t^\theta}{dQ_t} = \exp\{(\theta, X_t) - tX(\theta)\}, t \geq 0, \theta \in \Theta,$$

where $\chi(\theta) = lnE_Q \exp\{(\theta, X_1)\}, \Theta = \{\theta \in R^m : \chi(\theta) < \infty\}$.

After simple calculations we find that

$$P(\alpha; I) = P^{\theta(\alpha; I)}, \theta(\alpha; I) = \sum_{j=1}^k \alpha_j \theta_j,$$

$$h_t(\alpha; I) = t(\sum_{j=1}^k \alpha_j \chi(\theta_j) - \chi(\theta(\alpha; I))),$$

and

$$L_t(\alpha; I) = \exp\{(\theta(\alpha; I), X_t) - t\chi(\theta(\alpha; I))\},$$

denoting $(\cdot, \cdot)$ the inner product in R^m. We remark, that the general theory of exponential processes is reviewed in [11].

Example 2 (m-dimensional counting processes). Assume, that

$$\delta_t^\theta = \int_0^t \lambda^\theta(s)ds, \lambda^\theta(t) = (\lambda_1^\theta(t), \ldots, \lambda_m^\theta(t)),$$

$B_t \equiv 0, \sqcap^\theta$ and $\sqcap$ are concentrated at the points

$$e_j = (\underbrace{0, \ldots, 0}_{j-1}, 1, 0, \ldots, 0), j = 1, \ldots, m, \sqcap^\theta([0,t] \times \{e_j\})$$

$$= \int_0^t \lambda_j^\theta(s)ds, \delta_t = t1_m, \sqcap([0,t] \times e_j) = t, j = 1, \ldots, m, t \geq 0,$$

and there exist constants $0 < \underline{\lambda}_j^\theta < \overline{\lambda}_j^\theta$ such that

$$\underline{\lambda}_j^\theta \leq \lambda_j^\theta(t) \leq \overline{\lambda}_j^\theta, t \geq 0, j = 1, \ldots, m.$$

From the known properties of the counting processes and the theorems 1 and 2 we have that

$$N_t(\alpha; I) = \sum_{\ell=1}^m \int_0^t (\prod_{j=1}^k (\lambda_\ell^{\theta_j}(s))^{\alpha_j} - 1)d(X_s^\ell - s), X_t = (X_t^1, \ldots, X_t^m),$$

$$h_t(\alpha; I) = \sum_{\ell=1}^m \int_0^t \varphi_\alpha(\lambda_{\ell,I}(s))ds, \lambda_{\ell,I}(t) = (\lambda_\ell^{\theta_1}(t), \ldots, \lambda_\ell^{\theta_k}(t)).$$

$P(\alpha; I)$ is the probability measure, corresponding to the m-dimensional counting process with the conditional intensity vector

$$\lambda(\alpha; I)(t) = (\prod_{j=1}^k (\lambda_1^{\theta_j}(t))^{\alpha_j}), \ldots, \prod_{j=1}^k (\lambda_m^{\theta_j}(t))^{\alpha_j}).$$

Example 3 (m-dimensional diffusion type processes). Assume, that

$$\delta_t^\theta = \int_0^t b^\theta(s)ds, b^\theta(t) = (b_1^\theta(t), \ldots, b_m^\theta(t)),$$

$$\beta_t^{j\ell} = t\delta_{j,\ell}, j, \ell = 1, \dots, m, \sqcap^\theta \equiv 0, \delta_t \equiv 0,$$

$\sqcap \equiv 0$ and the functions $b^{\theta_j}(t), j = 1, \dots, m$, are uniformly bounded; $\delta_{j\ell}$ denotes the Kronecker symbols.

Denoting

$$b(\alpha; I)(t) = \sum_{j=1}^{k} \alpha_j b^{\theta_j}(t), t \geq 0,$$

and applying theorems 1 and 2 we find, that

$$N_t(\alpha; I) = \int_0^t b(\alpha; I)(s) dX_s,$$

$$h_t(\alpha; I) = \frac{1}{2} \int_0^t (\sum_{j=1}^{k} \alpha_j |b^{\theta_j}(s)|^2 - |b(\alpha; I)(s)|^2) ds.$$

$P(\alpha; I)$ is the probability measure, corresponding to the diffusion type process with the drift vector $b(\alpha; I)$ and the unit matrix of diffusion coefficients.

Assuming that $b^\theta(t) = b^\theta(X(t))$, where $b^\theta(x), x \in R^m, \theta \in \Theta$, are nonrandom and sufficiently smooth functions in x and the initial distributions are Dirac measures $\varepsilon_x(\cdot), x \in R^m$, we have that for each α and I the corresponding Hellinger measures $H_{t,x}(\alpha; I)(A), A \in \mathcal{F}_t$, generate the Schrödinger semigroup of linear operators

$$T_t(\alpha; I)(f)(x) = E_{p_{t,x}(\alpha; I)} e^{-h_t(\alpha; I)} f(X_t),$$

$$f \in C(R^m), t \geq 0, x \in R^m.$$

Denoting

$$b(\alpha; I)(x) = \sum_{j=1}^{k} \alpha_j b^{\theta_j}(x)$$

and

$$u(t, x) = T_t(\alpha; I)(f)(x);$$

we obtain, that

$$\frac{\partial u(t,x)}{\partial t} = \frac{1}{2}\Delta u(t,x) + (b(\alpha;I)(x), \nabla u(t,x)) -$$

$$-\frac{1}{2}(\sum_{j=1}^{k} \alpha_j |b^{\theta_j}(x)|^2 - |b(\alpha;I)(x)|^2) u(t,x),$$

$$u(0,x) = f(x).$$

If we denote $T_t(\theta), t \geq 0, \theta \in \Theta$, the semigroup of linear operators, generated by the diffusion process with the drift vector $b^{\theta}(x)$ and the unit diffusion matrix, then we can write symbolically that

$$T_t(\alpha; I) = \prod_{j=1}^{k} T_t(\theta_j)^{\alpha_j} \ (cf.[7]).$$

Finally we remark, that the domination assumption (1) in the theorems 1 and 2 can be relaxed, but not removed. Necessary and sufficient conditions are found in [7] for the probability measures, corresponding to the quasi-left continuous processes with the independent increments ($m = 1, k = 2, \alpha_1 = \alpha_2 = \frac{1}{2}$).

References

[1] S. Kakutani, On equivalence of infinite product measures, Ann. Math., 1948, vol. 49, p. 214-224.

[2] J. Jacod, A.N. Shiryaev, Limit theorems for stochastic processes, Springer, Berlin Heidelberg New York, 1987.

[3] F. Liese, I. Vajda, Convex statistical distances, Teubner-Texte zur Mathematik, B. 95, Leipzig, 1987.

[4] J. Jacod. Une application de la topologie d'Emery: le processus information d'un modéle statistique filtré, Sém. Probab. XXIII, Lecture Notes in Math., vol. 1372, Springer, Berlin Heidelberg New York, 1989, p. 448-474.

[5] I.A. Ibragimov, R.Z. Hasminski, Statistical estimation: asymptotic theory, Springer, Berlin Heidelberg New York, 1981.

[6] I. Vajda, Theory of statistical inference and information, Reidel, Dordrecht, 1988.

[7] J. Oosterhoff, W.r. Van Zwet, A note on contiguity and Hellinger distance, Contributions to statistics, ed. J. Jurečkova, Reidel, Dortrecht, 1979.

[8] C.M. Newman, The inner product of path space measures corresponding to random processes with independent increments, Bull. Amer. Math. Soc., 1972, vol. 78, No. 2, p. 268-272.

[9] J. Jacod, Filtered statistical models and Hellinger processes, Stochastic Proc. Appl., 1989, vol. 32, p. 3-45.

[10] R.S. Liptser, A.N. Shiryaev, Theory of martingales, Nauka, Moscow, 1986.

[11] U. Küchler, M. Sorensen, Exponential families of stochastic processes: a unifying semimartingale approach, Intern. Statisti. Rev., 1989, vol. 57, No. 2, p. 123-144.

International Series of Numerical Mathematics, Vol. 102, © 1991 Birkhäuser Verlag Basel

LATTICE APPROXIMATION OF A NONLINEAR STOCHASTIC PARTIAL DIFFERENTIAL EQUATION WITH WHITE NOISE

Gottfried Jetschke

Mathematische Fakultät
Friedrich-Schiller-Universität Jena
D-(O)-6900 Jena

Summary: A lattice approximation scheme in space of a nonlinear stochastic partial differential equation with right hand white noise is formulated. It is shown that convergence in probability takes place uniformly with respect to the space variable. Implications for the convergence of the distributions and interpretations of the physicist's approach as well as a suggestion for the problem of Simulated Annealing are given.

1. The model

Spatial transport and local production of a chemical substance as well as other similar phenomena usually are described by the following nonlinear reaction-diffusion equation ([3], [6], [2]):

$$\frac{\partial u}{\partial t}(t,x) = \frac{\partial^2 u}{\partial x^2}(t,x) + f[u(t,x)] \qquad t>0 , \quad x\in(0,L) ,$$
$$u(0,x) = h(x) . \tag{1}$$

We only consider a spatially one-dimensional situation, i.e. a long thin tube of length L. For simplicity we impose at the boundaries $x=0$ and $x=L$ homogeneous Dirichlet conditions,

$$u(t,0) = u(t,L) = 0 \tag{2}$$

Let us assume that the reaction function $f\colon \mathbb{R} \to \mathbb{R}$ is continuous and satisfies the following condition:

(A1) *There is a* $z>0$ *such that* $f(-u)>0$ *and* $f(u)<0$ *for all* $u>z$.

This condition ensures that the integral

$$F(u) := \int_0^u f(v)dv \tag{3}$$

is bounded above. A typical example is the cubic nonlinearity

$$f(u) = -u^3 + ru \quad , \quad F(u) = -(1/4)u^4 + (r/2)u^2 \le r^2/4 \quad . \tag{4}$$

With the identification $u_t \equiv u(t,\cdot)$ the deterministic system is understood as an evolution in a suitable function space. It is easy to verify that equation (1) then can be written as a gradient system

$$\frac{d}{dt}u_t = -\frac{\delta S}{\delta q}(u_t) \quad , \quad u_0 = h \tag{5}$$

with the "potential" or "Ginzburg-Landau functional"

$$S(q) := \int_0^L \left(\frac{1}{2}|q'(x)|^2 - F[q(x)]\right)dx \tag{6}$$

(where $\delta S/\delta q$ denotes the Frechet derivative).
Qualitative properties of the flow generated by (5) have been discussed in [9], [3], [1]. Using S as a Ljapunov functional, from the gradient dynamics (5) it follows $(d/dt)S(u_t) \leq 0$, hence a unique global solution $(u_t)_{t\geq 0}$ exists for all initial conditions h. Moreover, the only indecomposable ω-limit sets of the flow (5) are fixed points u^0 being stationary points of S, $(\delta S/\delta u)(u^0) = 0$.

The reaction-diffusion eqation (1),(2) can only describe the macroscopic behaviour of the thermodynamical or biological system. Since this consists of many particles or individuals in large but finite systems necessarily fluctuations occur which are usually modelled on a "mesoscopic" level by adding ad hoc a spatio-temporal white noise as a random source term. Hence the state becomes a random field U(t,x), and we arrive at the nonlinear stochastic partial differential equation ([6], [7], [15])

$$\frac{\partial U}{\partial t}(t,x) = \frac{\partial^2 U}{\partial x^2}(t,x) + f[U(t,x)] + \sigma\cdot\xi(t,x)$$
$$t \in (0,T] \quad , \quad x \in (0,L) \quad , \quad \sigma \geq 0 \quad , \tag{7a}$$

which should (for example) satisfy the homogeneous Dirichlet boundary conditions

$$U(t,0) = 0 \quad , \quad U(t,L) = 0 \qquad t \geq 0 \tag{7b}$$

and the initial condition

$$U(0,x) = h(x) \qquad x \in [0,L] \quad . \tag{7c}$$

Since the noise ξ should be extremely short-correlated it is assumed to be a Gaussian white noise in time and space, i.e. a formal Gaussian random field with

$$E\xi(t,x) = 0 \quad , \quad E\xi(t,x)\cdot\xi(t',x') = \delta(t-t')\cdot\delta(x-x') \quad . \tag{8}$$

The parameter σ measures the intensity of the noise.
From the nature of the problem it seems to be advantageous to consider problem (7) as a Markovian random evolution $U \equiv (U(t,\cdot))_{t\geq 0}$ in the space C[0,L] of functions over the spatial variable x. On the other hand it is often useful (and equivalent, see [10]) to consider $U \equiv (U(t,x))_{(t,x)\in[0,T]\times[0,L]}$ as a two-parameter random field.

In physicist's literature the transition probability of the MARKOV process (U_t) and its one-dimensional distributions are characterized by a formal density with respect to the (non-existing!) infinite-dimensional Lebesgue measure $d^\infty q$. They should satisfy the functional Fokker-Planck equation ([6], [2])

$$\frac{\partial}{\partial t}p(q,t) = -\frac{\delta}{\delta q}([\Delta q + f(q)]\cdot p(q,t)) + \frac{\sigma^2}{2}\cdot\frac{\delta^2 p}{\delta q^2} \quad . \tag{9}$$

Especially the density $p^o(q)$ of the stationary distribution can be calculated explicitely and is given by

$$p^o(q) = \tilde{N}\cdot\exp(-\frac{2}{\sigma^2}S(q)) \ , \quad S(q) = \int_0^L (\frac{1}{2}|q'(x)|^2 - F[q(x)])dx \tag{10}$$

with $F' \equiv f$ and $\tilde{N}$ as a normalization factor.

Equation (7) has, of course, only a formal meaning since the noise term ξ due to (8) is a generalized random field. A rigorous formulation of the underlying model can be given as follows ([10], [5]):

Let $(W_{tx})_{t\geq 0, x\geq 0}$ be a two-parameter Wiener field, i.e. a Gaussian centered field with $EW_{tx}W_{t'x'} = \min(t,t')\cdot\min(x,x')$, and choose for $(\mathcal{F}_t)_{t\geq 0}$ the family of σ-fields generated by $(W_{sx})_{(s,x)\in[0,t]\times[0,L]}$ and by $(h(x))_{x\in[0,L]}$, if the initial value h is random.

A solution of (7) is, by definition, a random field $U(t,x,\omega)$ which is $\mathcal{F}_t$-measurable for given t and all x and satisfies the stochastic integral equation

$$U_{tx}(\omega) = \int_0^L G(t,x,y)h(y,\omega)dy + \int_0^t\int_0^L G(t-s,x,y)f[U_{sy}(\omega)]dyds + \sigma\cdot M_{tx} \tag{11}$$

with

$$M_{tx} := \int_0^t\int_0^L G(t-s,x,y)dW_{sy}(\omega) \quad , \ (t,x)\in[0,T]\times[0,L] \ , \tag{12}$$

Here $G(t,x,y)$ is the fundamental solution of the corresponding linear deterministic problem, that is the heat equation (with Dirichlet boundary conditions (2)),

$$G(t,x,y) = \sum_{n=1}^{\infty} e^{-\lambda_n t}\cdot e_n(x)\cdot e_n(y) \tag{13}$$

with

$$\lambda_n = (n\pi/L)^2 \quad , \quad e_n(x) = \sqrt{2/L}\cdot\sin(n\pi x/L) \quad , \ n=1,2,\ldots \tag{14}$$

The term M_{tx} is a stochastic integral with respect to the two-parameter Wiener field W and constitutes a centered Gaussian random field with covariance

$$EM_{tx}M_{t'x'} = \int_0^{\min(t,t')} G(t+t'-2s,x,x')ds \tag{15}$$

and has a version with sample paths being continuous in (t,x). Hence, given (M_{tx}), the equation (11) can be solved pathwise in the subspace of $C([0,T]\times[0,L])$ satisfying the boundary conditions.

Alternatively the dynamical system (11) can be understood as an evolution in a suitable space of functions. Then the solution process (U_t) should fulfill the stochastic integral equation

$$U_t = T_t h + \int_0^t T_{t-s} f(U_s)ds + \sigma\cdot\int_0^t T_{t-s}dZ_s \quad , \quad t\in[0,T] \tag{16}$$

with (T_t) as the semigroup generated by $\Delta \equiv d^2/dx^2$ (together with the boundary conditions),

$$T_t = \sum_{k=1}^{\infty} e^{-\lambda_k t}\cdot e_k \circ e_k \quad , \tag{17}$$

and

$$M_t := \int_0^t T_{t-s}dZ_s \quad , \tag{18}$$

as the stochastic integral with respect to the infinite-dimensional cylindrical Wiener process (Z_t). It follows that M_t takes values in the space of continuous functions

$$C_{oo}[0,L] := \{q \mid q \in C[0,L] \ , \ q(0) = q(L) = 0\}$$

satisfying the boundary conditions (2). If the components Z_t^k of the cylindrical process (Z_t), being independent standard Wiener processes, are constructed from the WIENER field (W_{tx}) by

$$Z_t^k := \int_0^t\int_0^L e_k(x)dW_{sx} \ , \ k=1,2,...$$

then in a previous paper ([10]) it was shown that $M_{t(\cdot)} = M_t$ P-a.s. and, consequently, $U_{t(\cdot)} = U_t$ P-a.s. for all $t\in[0,T]$.

Conditions for existence and uniqueness of the solution of the non-linear equation are given in [16]. Suppose that the function $f:\mathbb{R}\to\mathbb{R}$ satisfies the following conditions:

(A2) f *is locally Lipschitz continuous* ,

(A3) *There exist nonincreasing functions* $g_1,g_2:\mathbb{R}\to\mathbb{R}$ *such that* $g_1(u) \le f(u) \le g_2(u)$ *for all* $u\in\mathbb{R}$.

Let h be a random variable taking a.s. values in the space $C_{oo}[0,L]$. Then for every $T>0$ equation (16) has a unique solution $(U_t)_{t\in[0,T]}$ with values in $C([0,T], C_{oo}[0,L])$.

2. Lattice approximation

Since the analytic treatment of spatially continuous system in the in stochastic case causes a lot of technical problems, in physicist's literature often a spatial "lattice" discretization is performed ([7],

[2], [18], [19]). This leads to a finite-dimensional problem with well-known techniques. We will show that this formal discretization can be formulated rigorously and gives good approximation properties of the paths. (Similar techniques have been used in [5], [8].)

2.1. Approximation scheme and reformulation

Usually the discretization divides the interval $[0,L]$ into N cells of length $l:=L/N$ and substitutes the second derivative d^2/dx^2 by a symmetric second difference. The noise $\xi(t,x)$ is replaced by independent white noise processes $\xi_i(t)$ at each lattice point, but the intensity has to be scaled by a factor $1/\sqrt{l}$.

Precisely formulated this model consists of the following system of nonlinear coupled stochastic differential equations:

$$\begin{aligned} dU_i(t) &= (\tfrac{1}{l^2}[U_{i+1}(t)+U_{i-1}(t)-2U_i(t)] + f[U_i(t)])dt + \tfrac{\sigma}{\sqrt{l}}dW_i(t) , \\ &\qquad t\geq 0 , \quad i=1,\dots,N-1 , \\ U_0(t) &= U_N(t) = 0 \quad \forall t\geq 0 , \quad U_i(0) = h_i , \quad i=1,\dots,N-1 , \end{aligned} \tag{19}$$

with the pointwise sampled intial value

$$h_i := h(il) \qquad i=1,\dots,N-1 .$$

The Wiener processes $(W_i(t))_{t\geq 0}$, $i=1,\dots,N-1$, are supposed to be independent and standard, i.e. $EW_i(t)W_k(s) = \delta_{ik}\cdot\min(t,s)$.

Theorem 1: *Let f satisfy (A1) - (A3). Then the system (19) has a unique global solution* $U_N := ((U_i(t))_{i=1}^{N-1})_{t\geq 0}$ *on* $\mathbb{R}^{N-1}$.

Remark: Since the boundary values are fixed we choose for convenience $\mathbb{R}^{N-1}$ to be the state space.

Proof: The application of standard methods (see f.e. [4], [17]) ensures the existence of a unique strong solution on a maximal time interval. Explosion can be excluded by another criterium ([20]). ∎

Now the system (19) shall be cast into a another form to allow a direct comparison with the continuous equation (11). Let $G_{ik}(t)$, $i,k=0,\dots,N+1$, $t\geq 0$, be the fundamental solution to the operator $d/dt-\Delta_N$ with

$$(\Delta_N u)_i := \frac{1}{l^2}(u_{i+1}+u_{i-1}-2u_i) , \quad u_0 = u_N = 0 . \tag{20}$$

We scale the initial condition by $1/l$, $G_{ik}(0) = (1/l)\cdot\delta_{ik}$, to give

$$\int_0^L G(0,x,y)dy = 1 = l\cdot\sum_{k=1}^{N-1} G_{ik}(0) , \quad x\in(0,L) , \quad i=1,\dots,N-1 .$$

Assertion: *The fundamental solution* $G_{ik}(t)$ *is given by*

$$G_{ik}(t) = \sum_{n=1}^{N-1} \exp(-\lambda_n^N t)\cdot e_n^N(i)\cdot e_n^N(k) \quad , \quad i,k=0,\dots,N \tag{21a}$$

with

$$\lambda_n^N = (4N^2/l^2)\cdot\sin^2(n\pi/2N) \quad , \quad e_n^N(j) = \sqrt{2/L}\cdot\sin(n\pi j/N) \quad , \quad j=0,\dots,N \ , \ n=1,\dots N-1 \quad . \tag{21b}$$

Proof: The fundamental solution can be obtained by calculating the eigenvalues λ_n^N and eigenvectors $e_n^N(\cdot)$, n=1,...N-1, of the matrix Δ_N. ∎

Remark: The vectors $e_n^N(\cdot)$, n=1,...,N-1 , form a complete orthonormal system in $\mathbb{R}^{N-1}$ if the scalar product is chosen by

$$(a,b) := l\cdot\sum_{n=1}^{N-1} a_k b_k \quad .$$

Theorem 3: *Every solution of (19) is also a solution of*

$$U_i(t) = l\cdot\sum_{k=1}^{N-1} G_{ik}(t)h_k + l\cdot\int_0^t \sum_{k=1}^{N-1} G_{ik}(t-s)f[U_k(s)]ds + \sigma\sqrt{l}\cdot\int_0^t \sum_{k=1}^{N-1} G_{ik}(t-s)dW_k(s) \qquad i=1,\dots,N-1 \ , \tag{22}$$

(with $U_0(t) = U_N(t) = 0$ *) and vice versa.*

Remark: The theorem states that in finite dimensional cases the notions of strong and mild solutions coincide.

Much more generally the following theorem can be proved.

Theorem 4: *Let be* $A\in\mathcal{L}(\mathbb{R}^n)$ *a linear operator and* $T_t := e^{tA}$, $t\in\mathbb{R}$, *the generated group, let f and g be measurable funktions, let h be a random vector in* $\mathbb{R}^n$, *and let W be an n-dimensional Wiener process. Suppose that the stochastic differential equation*

$$U_t = h + \int_0^t [AU_s + f(U_s)]ds + \int_0^t g(U_s)dW_s \quad , \quad t\in[0,T] \tag{23}$$

has a unique solution. Then (U_t) *also solves the integral equation*

$$U_t = T_t h + \int_0^t T_{t-s} f(U_s)ds + \int_0^t g(U_s)dW_s \quad , \quad t\in[0,T] \tag{24}$$

and vice versa.

Proof: If (U_t) is a solution of (23) then its stochastic differential is

$$dU_s = [AU_s + f(U_s)]ds + g(U_s)dW_s \quad , \quad U_0 = h \quad . \tag{25}$$

According to the Ito formula the differential of $X_s := T_{t-s}U_s$ is

$$dX_s = T_{t-s}f(U_s)ds + T_{t-s}g(U_s)dW_s \quad , \tag{26}$$

where we used the relation $\partial T_{t-s}/\partial s = -AT_{t-s} = -T_{t-s}A$. This gives

$$U_t - T_t h = X_t - X_0 = \int_0^t T_{t-s}f(U_s)ds + \int_0^t T_{t-s}g(U_s)dW_s \quad ,$$

i.e. (U_t) also solves (24). Otherwise let (U_s) be a solution of (24). Then (X_t) has the stochastic differential (26). Since (T_t) is a group $U_s = T^{-1}_{t-s}X_s = T_{s-t}X_s$ has the differential (25), where we now have to use $\partial T_{s-t}/\partial s = AT_{s-t}$. Hence U also solves (23). ∎

Proof of Theorem 3: Put $n=N-1$, $g(u)\equiv\sigma/\sqrt{l}$ und $A=\Delta_N$ to obtain

$$(T_t u)_i \equiv (e^{tA}u)_i = l\cdot\sum_{k=1}^{N-1} G_{ik}(t)u_k \quad .$$

Then Theorem 3 is a special case of Theorem 4. ∎

To compare with the solution (U_{tx}) of the spatially continuous problem (11) the state space $\mathbb{R}^{N-1}$ of the problem (22) has to be embedded into the space $C_{oo}[0,L]$ of continuous functions on $[0,L]$ with zero boundary values. This can be done most easily by identifying $q\in\mathbb{R}^{N-1}$ with a piecewise linear function $\hat{q}$, i.e. by linear interpolation in x ,

$$L_N:\ (q_i)_{i=1}^N \mapsto \hat{q}_N(x) := \sum_{i=0}^{N-1} [q_i + (q_{i+1}-q_i)(\tfrac{x}{l} - i)]\cdot 1_{[il,(i+1)l]}(x) \quad , \tag{27}$$

(with $q_0=q_N=0$ implying $\hat{q}_N(0)=\hat{q}_N(L)=0$). Obviously the map L_N is one-to-one (for fixed N).

Definition: Let (U_{tx}), $(t,x)\in[0,T]\times[0,L]$, be the solution of the stochastic PDE (11) and let $(U_i(t))_{i=1}^{N-1}$, $t\geq 0$, be the solution of the (N-1)-dimensional stochastic ODE (19). Then

$$U^N_{tx} := L_N[(U_i(t))_{i=1}^{N-1}] \quad , \quad (t,x)\in[0,T]\times[0,L] \tag{28}$$

is called finite-dimensional (or lattice) approximation of (U_{tx}).

Let S_N be the operator, creating from any $(q_i)_{i=1}^{N-1}$ a step function q_N being right continuous in x ,

$$S_N:\ (q_i)_{i=1}^{N-1} \mapsto q_N(x) := \sum_{i=0}^{N-1} q_i\cdot 1_{[il,(i+1)l)}(x) + q_N\cdot 1_{\{L\}}(x) \quad , \tag{29}$$

(again with $q_0=q_N=0$ implying $q_N(0)=q_N(L)=0$).

Without changing the notation we will apply the operators L_N and S_N also to functions $q(x)$ of a continuous argument x. This is understood by acting on the sampled values $q(il)$, $i=1,\ldots,N-1$.

Let $G_N(t,x,y)$ be the function constructed from $G_{ik}(t)$ by linear interpolation in x and piecewise constant continuation in y ,

$$G_N(t,x,y) \quad := \quad L_N[S_N(G_{ik}(t))_{k=1}^{N-1}]_{i=1}^{N-1} \quad , \tag{30}$$

where S_N acts on k and L_N on i. The "Green's function" G_N is then derived from (21) by

$$G_N(t,x,y) \quad = \sum_{n=1}^{N-1} \exp(-\lambda_n^N t)\cdot\hat{e}_n^N(x)\cdot e_n^N(y) \tag{31a}$$

with

$$\lambda_n^N = (4N^2/l^2)\sin^2(n\pi/2N) \ , \quad \hat{e}_n^N(x) = L_N(e_n^N(i)) \ , \quad e_n^N(x) = S_N(e_n^N(i))$$

and

$$e_n^N(i) = \sqrt{2/L}\cdot\sin(n\pi i/N) \quad , \quad i=1,\dots,N \quad . \tag{31b}$$

(Notice that even $\hat{e}_n^N = L_N e_n$, $e_n^N = S_N e_n$, $n=1,\dots,N-1$, for the eigenfunctions e_n of Δ from (14) holds !)

Moreover, for any function $q \in C_{oo}[0,L]$, especially for any polygonial line, we put

$$f_N := f\circ S_N \tag{32}$$

in the sense of $f_N[q(x)] := f[S_N(q(x)] = f[q(il)]$ for all x from $[il,(i+1)l)$ and with $f_N[q(L)]=f(0)$.

Theorem 5: *For suitably chosen Wiener processes in (19) the lattice approximation (28) satisfies the stochastic integral equation*

$$U_{tx}^N = \int_0^L G_N(t,x,y)h_N(y)dy + \int_0^t\int_0^L G_N(t-s,x,y)f_N(U_{sy}^N)dyds + \sigma\cdot M_{tx}^N \tag{33}$$

with

$$M_{tx}^N := \sigma\int_0^t\int_0^L G_N(t-s,x,y)dW_{sy} \quad , \quad (t,x)\in[0,T]\times[0,L] \ , \tag{34}$$

where (W_{tx}) *is a two-parameter Wiener field and* $h_N := S_N h$ *the stepwise approximation of the initial value* h.

Proof: From the definition (30) of G_N and from $h_N = S_N h$ we obtain

$$\int_0^L G_N(t,il,y)h_N(y)dy \quad = \quad l\cdot\sum_{k=1}^{N-1} G_{ik}(t)h_k \quad , \quad i=1,\dots,N-1 \quad .$$

The definition (32) of f_N with $f_N(U_{sy}^N) = f(U_{s,kl}^N)$, $y\in[kl,(k+1)l)$, gives

$$\int_0^L G_N(t-s,il,y)f_N(U_{sy}^N)dy = l\cdot\sum_{k=1}^{N-1} G_{ik}(t-s)f[U_k(s)] \quad , \ i=1,\dots,N-1.$$

Since $G_N(t,x,y)$ is piecewise constant in y , it is easy to show

$$\int_0^t\int_0^L G_N(t-s,x,y)dW_{sy} \quad = \quad \sqrt{l}\cdot\sum_{k=1}^{N-1} G_N(t-s,x,kl)dW_s^k \quad ,$$

where the Wiener processes $(W_t^k)_{t\geq 0}$, $k=1,\dots,N-1$, are constructed by

$$W_t^k := \frac{1}{\sqrt{l}}\cdot\int_0^t\int_0^L 1_{[kl,(k+1)l)}(y)dW_{sy} \quad , \quad k=1,\dots,N-1 \tag{35}$$

from the Wiener field W and turn out to be independent. That means: If the N-1 Wiener processes in (19) (or (22)) are chosen according to (35) then especially it holds

$$\sigma\cdot\int_0^t\int_0^L G_N(t-s,il,y)dW_{sy} = \sigma\cdot\sqrt{l}\cdot\int_0^t\sum_{k=1}^{N-1} G_{ik}(t-s)dW_s^k \qquad P\text{-a.s.} \quad .$$

Hence the solution (U^N_{tx}) of (33), being polygonial in x, coincides at $x=il$, $i=1,\ldots,N-1$, for all $t\geq 0$ with $(U_i(t))$, $i=1,\ldots,N-1$, from (22) and satisfies the homogeneous boundary conditions. ∎

2.2. Uniform convergence of the stochastic integrals

In this section we will prove for the sequence of "noise terms" (M^N_{tx}) the uniform convergence in probability, which implies the almost sure convergence of a suitable subsequence. For continuous functions $u\in C([0,T]\times[0,L])$ of space and time we denote the maximum norm by

$$\|u\|_T := \sup_{(t,x)\in[0,T]\times[0,L]} |u_{tx}| \equiv \sup_{t\in[0,T]} \|u_{t(.)}\| \quad ,$$

where $\|\cdot\|$ is the maximum norm in $C[0,L]$.

The proof uses the fact that for the distributions P_{M^N-M} of the difference of the processes M^N und M the weak convergence to a one-point distribution concentrated at the zero function can be shown, which implies the uniform convergence in probability. Since the fields are Gaussian we have to show the weak convergence of the finite-dimensional distributions and the tightness of the system of measures.

Lemma 1: *For the eigenvalues* $\lambda_n=(n\pi/L)^2$, $n=1,2,\ldots$, *from (14) and all* $a>0$ *it holds*

$$\sum_{n=1}^{\infty} \min\,(a^2,\ 1/\lambda_n) \leq \frac{2L}{\pi}\cdot a \quad .$$

Proof: Let $[x]$ denote the integer part of a real variable x. Obviously $a^2\leq 1/\lambda_n = (l/n\pi)^2$ holds for exactly one $n<L/\pi a$. This implies

$$\sum_{n=1}^{\infty} \min\,(a^2,\ 1/\lambda_n) = \sum_{n=1}^{[L/\pi a]} a^2 + \sum_{n=[L/\pi a]+1}^{\infty} 1/\lambda_n \leq$$

$$\leq a^2\cdot\frac{L}{\pi a} + \frac{L^2}{\pi^2}\cdot\int_{[L/\pi a]}^{\infty} t^{-2}dt = \frac{2La}{\pi} \quad .$$

∎

Lemma 2: *For all* $N=1,2,\ldots$ *and all* $n=1,\ldots,N-1$ *the eigenvalues* λ_n *and* λ^N_n *satisfy*

$$(4/\pi^2)\cdot\lambda_n \leq \lambda^N_n \leq \lambda_n$$

and

$$0 \le (1/\lambda_n) + (1/\lambda_n^N) - 4/(\lambda_n+\lambda_n^N) \le [1-(\lambda_n^N/\lambda_n)]^2/\lambda_n^N \quad .$$

Proof: According to (14) and (21) we find

$$\lambda_n^N/\lambda_n = [\frac{\sin(n\pi/2N)}{(n\pi/2N)}]^2 \quad .$$

For $n\in\{1,\ldots,N-1\}$ the argument $n\pi/2N$ is from the interval $[0,\pi/2]$. Due to the unequality

$$1 \le \frac{\sin x}{x} \le \frac{\pi}{2}$$

the first part of the lemma follows. Morover, for $0 < c < d$ it holds

$$(1/d) + (1/c) - 4/(c+d) = (d-c)^2/cd(c+d) \le [1-(c/d)]^2/c \quad ,$$

which proves the second part.

■

Now we have to consider some properties of the stochastic integrals

$$M_{tx}^N := \int_0^t\int_0^L G_N(t-s,x,y)dW_{sy} \;, \quad M_{tx} = \int_0^t\int_0^L G(t-s,x,y)dW_{sy} \;. \quad (36)$$

Lemma 3: *For the random fields* $X=M$ *and* $X=M^N$ *in (36) the estimations*

$$E|X_{tx}-X_{ty}|^2 \le \frac{\pi\sqrt{L}}{2}\cdot\sqrt{|x-y|} \quad , \quad E|X_{tx}-X_{sx}|^2 \le \pi\sqrt{2}\cdot\sqrt{|t-s|}$$

are valid for all $t,s\in[0,T]$ *and all* $x,y\in[0,L]$.

Proof: Since M_{tx}^N is a stochastic integral we obtain

$$E|M_{tx}^N - M_{ty}^N|^2 = \int_0^t\int_0^L |G_N(t-s,x,z) - G_N(t-s,y,z)|^2 dzds =$$

$$= \int_0^t\int_0^L \Big| \sum_{n=1}^{\infty} e^{-\lambda_n^N(t-s)}\cdot[\hat{e}_n^N(x) - \hat{e}_n^N(y)]\cdot e_n^N(z)\Big|^2 dzds \quad .$$

The integrand is piecewise constant in z and the eigenvectors $e_n^N(\cdot)$ from (31) are orthonormal in $\mathbb{R}^{N-1}$, hence we get

$$E|M_{tx}^N - M_{ty}^N|^2 = \sum_{n=1}^{N-1} \int_0^t e^{-2\lambda_n^N(t-s)} ds \cdot |\hat{e}_n^N(x) - \hat{e}_n^N(y)|^2 \le$$

$$\le \sum_{n=1}^{N-1} \frac{1}{2\lambda_n^N}\cdot\frac{2}{L}\min((\frac{n\pi|x-y|}{L})^2, 1) \quad .$$

Lemma 1 and the estimation $1/\lambda_n^N < (\pi/2)^2\cdot(1/\lambda_n)$ from Lemma 2 give

$$E|M_{tx}^N - M_{ty}^N|^2 \le \frac{\pi^2}{4L}\cdot\sum_{n=1}^{\infty}\min(|x-y|^2, 1/\lambda_n) \le \frac{\pi}{2}\cdot|x-y| \quad .$$

Since $|x-y|\le\sqrt{L\cdot|x-y|}$ the first part of the lemma follows.
In a similar way we obtain for the other difference

$$E|M^N_{tx} - M^N_{sx}|^2 = E\left|\int_0^t\int_0^L G_N(t-r,x,y)dW_{ry} - \int_0^s\int_0^L G_N(s-r,x,y)dW_{ry}\right|^2 =$$

$$= \int_0^s\int_0^L |G_N(t-r,x,y) - G_N(s-r,x,y)|^2 dydr + \int_s^t\int_0^L |G_N(s-r,x,y)|^2 dydr =$$

$$= \sum_{n=1}^{N-1} |\hat{e}^N_n(x)|^2 \cdot \Big(\int_0^s \left| e^{-\lambda^N_n(t-r)} - e^{-\lambda^N_n(s-r)} \right|^2 ds + \int_s^t e^{-2\lambda^N_n(t-r)} ds \Big) .$$

Together with $(1-u)^2 \leq 1-u^2$ for $0 \leq u \leq 1$ calculation of the integrals yields

$$E|M^N_{tx} - M^N_{sx}|^2 \leq \frac{2}{L}\sum_{n=1}^{N-1} \frac{1}{\lambda^N_n}(1 - e^{-2\lambda^N_n(t-s)}) \quad .$$

The lower and upper estimations for λ^N_n from Lemma 2 , the unequality $1-\exp(-u) \leq \min(u,1)$ and Lemma 1 give then

$$E|M^N_{tx} - M^N_{sx}|^2 \leq \frac{\pi^2}{2L}\sum_{n=1}^{\infty} \min(2|t-s|, \frac{1}{\lambda_n}) \leq \pi\sqrt{2}\cdot\sqrt{|t-s|} \quad ,$$

and this proves the second part of the lemma for the field $X=M^N$.

For the field $X=M$ similar calculations yield

$$E|M_{tx} - M_{ty}|^2 = \int_0^t\int_0^L \left| \sum_{n=1}^{\infty} e^{-\lambda_n(t-s)} [e_n(x) - e_n(y)]\cdot e_n(z) \right|^2 dzds$$

$$= \sum_{n=1}^{\infty} \int_0^t e^{-2\lambda_n s} ds \cdot |e_n(x) - e_n(y)|^2 \leq$$

$$\leq \frac{2}{L}\sum_{n=1}^{\infty} \frac{1}{2\lambda_n}\cdot\min((\frac{n\pi|x-y|}{L})^2, 1) = \frac{1}{L}\sum_{n=1}^{\infty} \min(|x-y|^2, \frac{1}{\lambda_n}) \quad .$$

Together with Lemma 1 we get finally

$$E|M_{tx} - M_{ty}|^2 \leq \frac{2}{\pi}\cdot|x-y| \leq \frac{\pi}{2}\cdot\sqrt{L}\cdot\sqrt{|x-y|} \quad .$$

In a similar way we obtain

$$E|M_{tx} - M_{sx}|^2 = \int_0^s\int_0^L \left|\sum_{n=1}^{\infty} (e^{-\lambda_n(t-r)} - e^{-\lambda_n(s-r)}) e_n(x)e_n(y)\right|^2 dydr$$

$$+ \int_s^t\int_0^L \left|\sum_{n=1}^{\infty} e^{-\lambda_n(t-r)} e_n(x)e_n(y)\right|^2 dydr \leq$$

$$\leq \frac{4}{L}\cdot\sum_{n=1}^{\infty} \frac{1}{2\lambda_n}(1 - e^{-2\lambda_n(t-s)}) \leq \frac{2}{L}\cdot\sum_{n=1}^{\infty} \frac{1}{\lambda_n}\cdot\min(2\lambda_n|t-s| , 1) \leq$$

$$\leq \frac{2}{L}\cdot\sum_{n=1}^{\infty} \min(2|t-s| , 1/\lambda_n) \leq \frac{4}{\pi}\cdot\sqrt{2|t-s|} \leq \pi\sqrt{2}\cdot\sqrt{|t-s|} \quad ,$$

hence the lemma is proved completely. ∎

Remark: Notice that the estimations in Lemma 3 do not depend on N.

Lemma 4: *The sequence of distributions* $(P_{M^N-M})_{N=1}^{\infty}$ *is tight as a family of probability measures on the Borel sets of* $C([0,T]\times[0,L])$.

Proof: The random fields M und M^N are continuous in t and x, hence it suffices to show that there are positive constants p , α und C , such that

$$\sup_N E|M^N_{tx} - M_{tx} - (M^N_{sy} - M_{sy})|^p \le C\cdot(|t-s|+|x-y|)^{2+\alpha}$$

holds for all $t,s\in[0,T]$ and all $x,y\in[0,L]$.

We can rearrange

$$|M^N_{tx} - M_{tx} - M^N_{sy} + M_{sy}|^2 \le (|M^N_{tx} - M^N_{sy}| + |M_{tx} - M_{sy}|)^2 \le$$

$$\le 2(|M^N_{tx} - M^N_{ty}| + |M^N_{ty} - M^N_{sy}|)^2 + 2(|M_{tx} - M_{ty}| + |M_{ty} - M_{sy}|)^2 \le$$

$$\le 4|M^N_{tx} - M^N_{ty}|^2 + 4|M^N_{ty} - M^N_{sy}|^2 + 4|M_{tx} - M_{ty}|^2 + 4|M_{ty} - M_{sy}|^2 \quad ,$$

hence we get together with Lemma 3

$$E|M^N_{tx} - M_{tx} - (M^N_{sy} - M_{sy})|^2 \le \text{const}\cdot(\sqrt{|t-s|} + \sqrt{|x-y|}) \quad ,$$

where the constant does not depend on T and especially not on N, but only depends on L. Taking into account the Gaussian nature of the field M^N-M we obtain finally

$$\sup_N E|M^N_{tx} - M_{tx} - (M^N_{sy} - M_{sy})|^{10} \le C\cdot(|t-s| + |x-y|)^{5/2} \quad ,$$

which proves the tightness of the distributions of M^N-M . ∎

Lemma 5: *For the random fields* M *and* M^N *from (36) we have*

$$\sup_{(t,x)\in[0,T]\times[0,L]} E|M^N_{tx} - M_{tx}|^2 \xrightarrow[N\to\infty]{} 0 \qquad (37)$$

Proof: From the definition of the stochastic integrals we obtain

$$E|M^N_{tx} - M_{tx}|^2 = \int_0^t\int_0^L |G(t-s,x,y) - G_N(t-s,x,y)|^2 dyds \quad .$$

Inserting the special structure of G and G_N from (13) and (31) and using a suitable decomposition and the triangle inequality the integral can be estimated above and shown to be convergent to zero in N (see [13], [5]). ∎

Remark: The left hand side in (37) can be estimated in N and we found [13]

$$\sup_{(t,x)\in[0,T]\times[0,L]} E|M^N_{tx} - M_{tx}|^2 \le \text{const}\cdot\frac{1}{N} \quad .$$

This implies

$$P(\sup_{(t,x)\in[0,T]\times[0,L]} |M^N_{tx} - M_{tx}| > \varepsilon) \le \frac{\text{const}}{\varepsilon^4}\cdot\sum_{N=1}^{\infty}\frac{1}{N^2} < \infty \quad .$$

so we have even proved

$$M^N_{tx} \underset{N\to\infty}{\longrightarrow} M_{tx} \qquad P\text{-a.s.} \qquad \forall (t,x)\in[0,T]\times[0,L] \quad ,$$

i.e. the P-almost sure pointwise convergence of the noise terms! Unfortunately this type of convergence cannot be transferred to the solutions of the full nonlinear eqation (only to the linear one).

Theorem 6: a) *For the random fields* (M^N_{tx}) *and* (M_{tx}) *in (36) it holds*

$$\lim_{N\to\infty} P\{\|M_{tx} - M^N_{tx}\|_T > \varepsilon\} = 0 \qquad \forall\varepsilon>0 \quad .$$

b) *For a suitable subsequence* (N_R) , R=1,2,... *we have*

$$\lim_{R\to\infty} \|M_{tx}(\omega) - M^{N_R}_{tx}(\omega)\|_T = 0 \qquad P\text{-a.s.} \quad .$$

Proof: From Lemma 5 we can conclude

$$[E(M^N_{tx} - M_{tx})(M^N_{sy} - M_{sy})]^2 \leq E|M^N_{tx} - M_{tx}|^2 \cdot E|M^N_{sy} - M_{sy}|^2 \underset{N\to\infty}{\longrightarrow} 0$$

for all $t,s\in[0,T]$ and for all $x,y\in[0,L]$. This implies that all finite dimensional distributions of the Gaussian field M^N-M as $N\to\infty$ converge weakly to a one-point distribution in zero. Together with the tightness from Lemma 4 this implies the weak convergence of the induced measures on the space $C([0,T]\times[0,L])$ to a one-point distribution on the function being identically zero,

$$P_{M^N-M} \underset{N\to\infty}{\Rightarrow} \delta_0 \quad .$$

But this immediately gives

$$\lim_{N\to\infty} P\{\|M^N_{tx} - M_{tx}\|_T > \varepsilon\} = 0 \qquad \forall\varepsilon>0 \quad ,$$

hence the uniform convergence of the noise terms in probability. For a suitable subseqence even the P-almost sure uniform convergence follows.

■

Remarks: 1) Since all estimations and constants do not depend on T , we have tightness, weak convergence and uniform convergence in probability also in the space $C([0,\infty)\times[0,L])$.

2) The same statement also holds for other boundary conditions.

3) It is of great practical interest to know explicitly a subsequence which indeed converges uniformly P-a.s. in N. For this purpose it is necessary to estimate how fast the sequence of probabilities $P\{\|M^N-M\|_T > \varepsilon\}$ decreases in N. Although the measures are Gaussian with a well-known covariance functional, these estimations are complicated and not known from the literature. It is suggested that a

doubling of the number of lattice points, $N_R = 2R$, should be sufficient.

(Even in the much simpler case of a family of one-dimensional Ornstein-Uhlenbeck processes (Y^a), $a>0$, all starting in zero,

$$Y_t^a := \int_0^t e^{-a(t-s)} dW_s \qquad , \quad t \geq 0 ,$$

it is easy to show

$$P(\sup_{0 \leq t \leq T} |Y_t^a| > \varepsilon) \underset{a \uparrow \infty}{\longrightarrow} 0 \qquad \text{for all } \varepsilon > 0 ,$$

hence (Y^a) converges uniformly to the zero process in probability as $a \to \infty$. But it is still an open problem to estimate this probability in its dependence on a. All known techniques (decomposition of Y^a into a martingale plus process of bounded variation, Girsanov formula etc.) failed up to now, since Y^a is neither a martingale nor a super- or submartingale.)

2.3. Uniform convergence of the solutions

The result of Theorem 6 remains true word by word for the solution U of the nonlinear equation (11) and its lattice approximation U^N from (33), since the dependence on M and M^N, respectively, is deterministic.

To prove this result we need a theorem, which states the continuous dependence of the solution U on the stochastic integral M (or, respectively, the solution V of the linear equation) and the same for the lattice case.

Let us denote by V the solution of the linear equation, i.e. (16) with $f \equiv 0$,

$$V_t = T_t h + \sigma \cdot M_t \qquad , \qquad t \in [0,T] \quad . \tag{38}$$

Theorem 5: *The solution* U *from (16) depends continuously on the linear solution V from (38), and the dependence is one-to-one.*

The proof is given in [13] and uses essentially the fact, that (16) has a unique solution with continuous dependence on h and M. Another proof was given in [1], showing at the same time that the solution is unique, but only for the special case of the cubic nonlinearity.

The theorem states that there is a one-to-one continuous map $U = \Psi(V)$, acting on $C([0,T] \times [0,L])$. Comparing (16) and (38) we see that the inverse $V = \Psi^{-1}(U)$ is given by

$$V_t = U_t - \int_0^t T_{t-s} f(U_s) ds \qquad , \quad t \in [0,T] \quad . \tag{39}$$

In a similar way we can prove that there is a unique one-to-one map Ψ^N on the same space connecting the polygon-type solutions V^N (with $f\equiv 0$) and U^N.

Theorem 7: a) *For the solution* $U=(U_{tx})$ *of equation (12) and the approximating solution* $U^N=(U^N_{tx})$ *from equation (33) we have*

$$\lim_{N\to\infty} P(\|U_{tx}-U^N_{tx}\|_T > \varepsilon) = 0 \qquad \forall\varepsilon>0 \quad .$$

b) *For a suitably chosen subsequence* (N_R), R=1,2,... *it holds*

$$\lim_{R\to\infty} \|U_{tx}(\omega) - U^{N_R}_{tx}(\omega)\|_T = 0 \qquad \text{P-f.s.} \quad .$$

Proof: *Step 1:* Let V and V^N be the solutions of the linear equations, i.e.

$$V_{tx} = \int_0^L G(t,x,y)h(y)dy + \sigma\cdot M_{tx} \quad , \tag{40}$$

$$V^N_{tx} = \int_0^L G_N(t,x,y)h_N(y)dy + \sigma\cdot M^N_{tx} \quad , \tag{41}$$

with $h_N(y)=h(kl)$ for $y\in[kl,(k+1)l)$, $k=0,\dots,N-1$, $h_N(L)=0$ and

$$U_{tx} = V_{tx} + \int_0^t\int_0^L G(t-s,x,y)f(U_{sy})dyds \quad , \tag{42}$$

$$U^N_{tx} = V^N_{tx} + \int_0^t\int_0^L G_N(t-s,x,y)f_N(U_{sy})dyds \quad . \tag{43}$$

From the representations (31) of G_N and (13) of G we can derive easily that for any $h\in C[0,L]$ it holds

$$\left\| \int_0^L G_N(\cdot,\cdot,y)h_N(y)dy - \int_0^L G(\cdot,\cdot,y)h(y)dy \right\|_T \xrightarrow[N\to\infty]{} 0 \quad . \tag{44}$$

Suppose that the random initial value h takes its value P-a.s. in $C[0,L]$. Then the difference of both integrals in (40) and (41) converges due to (44) P-a.s. to zero. This implies

$$\|V_{tx}-V^N_{tx}\|_T \xrightarrow[N\to\infty]{} 0 \qquad \text{in probability} \tag{45}$$

and

$$\|V_{tx}-V^{N_R}_{tx}\|_T \xrightarrow[R\to\infty]{} 0 \qquad \text{P-a.s.} \tag{46}$$

for a suitable subsequence.

Step 2: The inverses $\Phi\equiv\Psi^{-1}$ and $\Phi_N\equiv\Psi_N^{-1}$ of the functions $U=\Psi(V)$ and $U^N=\Psi_N(V^N)$ are given by

$$\Phi(u)_t = u_t - \int_0^t T_{t-s}f(u_s)ds \quad , \qquad \Phi_N(u)_t = u_t - \int_0^t T^N_{t-s}f_N(u_s)ds \; ,$$

with (T_t^N) as the semigroup generated by Δ_N ,

$$T_{t-s}^N f_N(u_s) := \int_0^L G_N(t-s,\cdot,y) f_N(u_{s(\cdot)})dy = l\cdot\sum_{k=1}^{N-1} G_N(t-s,\cdot,kl) f(u_{s,kl}) \ .$$

The last representation, obtained by using the definition (32) of f_N , shows that Φ_N can be defined on all continuous functions.

Our aim is now to show that for all $\varepsilon>0$ there can be found an $\varepsilon_1>0$ such that for all $N\geq N_o$ the implication

$$\|V-V^N\|_T \leq \varepsilon_1 \quad \longrightarrow \quad \|U-U^N\|_T \leq \varepsilon$$

is valid. (This is a deterministic property of the maps Ψ and Ψ_N !) With this the assertions (45) and (46) on convergence can be carried over to U und U^N .

Let $K_\delta(U) := \{u \mid \|u-U\|_T < \delta\}$ be an open ball of radius δ around U in the space $C([0,T]\times[0,L])$ and let

$$A_N(U) := \{v \mid v=\Phi_N(u) \ , \ u\in K_\delta(U)\}$$

be the images of K_δ under the maps Φ_N . Next we will show that there is an open ball around $V=\Phi(U)$, which is contained in all sets $A_N(U)$. Due to the one-to-one property its image under Ψ_N must be contained in $K_\delta(U)$. Then for $U^N\in K_\delta(U)$ it holds

$$|U_t - U_t^N| = |V_t - V_t^N + \int_0^t [T_{t-s}f(U_s) - T_{t-s}^N f_N(U_s^N)]ds \ | \leq$$

$$\leq |V_t - V_t^N| + \int_0^t |T_{t-s}f(U_s) - T_{t-s}^N f_N(U_s)|ds + \int_0^t T_{t-s}^N |f_N(U_s) - f_N(U_s^N)|ds \ .$$

Because of (44) we have

$$\|T_{t-s}f(U_s) - T_{t-s}^N f_N(U_s)\| \leq \varepsilon_o \qquad \text{for } N\geq N_o \quad ,$$

and since $U^N\in K_\delta(U)$ it holds

$$|f_N(U_s) - f_N(U_s^N)| \leq L\cdot |U_s - U_s^N| \quad ,$$

where $L=L(\delta,U)$ is a suitable local Lipschitz constant. Moreover, since $(T_t^N)_{t\in\mathbb{R}}$ is a family of contractions,

$$\|U_t - U_t^N\| \leq \|V_t - V_t^N\| + \varepsilon_o T + L\cdot\int_0^t \|U_s - U_s^N\|ds \quad ,$$

and Gronwall's lemma gives

$$\|U - U^N\|_T \leq e^{LT}(\|V - V^N\|_T + \varepsilon_o T) \qquad \forall N\geq N_o \ . \tag{47}$$

For all U^N on $\partial K_\delta(U)$, that is for $\|U-U^N\|=\delta$, finally we find

$$\|V - V^N\|_T \geq \delta\cdot e^{-LT} - \varepsilon_o T \qquad \forall N\geq N_o \ .$$

Now we choose ε_o so small (and therefore N_o so large) that

$$\delta\cdot e^{-LT} - \varepsilon_o T \geq d > \varepsilon_o T > 0 .$$

Then for all $U^N \in \partial K_\delta(U)$ it holds

$$\|\Phi(U) - \Phi_N(U^N)\|_T = \|V - V^N\|_T \geq d > 0 \qquad \forall N \geq N_o .$$

Step 3: Since Φ_N is continuous and one-to-one it holds

$$\Phi_N(U) \in \Phi_N[K_\delta(U)] .$$

For $\Phi_N(U)$ and $V=\Phi(U)$ we get

$$\Phi_N(U)_t - V_t = \int_0^t [T_{t-s} f(U_s) - T^N_{t-s} f_N(U_s)]ds ,$$

i.e.

$$\|\Phi_N(U) - V\|_T \leq \varepsilon_o T \qquad \forall N \geq N_o .$$

Suppose that $V \notin \Phi_N[K_\delta(U)]$, then

$$\|\Phi_N(U) - V\|_T \geq \inf_{V^N \in \Phi_N[\partial K_\delta(U)]} \|V^N - V\| \geq d \qquad \text{for } N \geq N_o$$

contradicts $\varepsilon_o T < d$. Therefore V is contained in $\Phi_N[K_\delta(U)]$ for all $N \geq N_o$. Hence the ball $\|V-V^N\| < d$ is contained in the intersection of all sets $\Phi_N[K_\delta(U)]$.

We choose $\varepsilon_1 < d$ and know

$$\bar{K}_{\varepsilon_1}(V) := \{v \mid \|v-V\|_T \leq \varepsilon_1\} \subset \bigcap_{N \geq N_o} \Phi_N[K_\delta(U)] .$$

All functions V^N with $\|V-V^N\| \leq \varepsilon_1$ therefore lead to

$$\|U - U^N\|_T = \|\Psi(V) - \Psi_N(V^N)\|_T \leq \delta .$$

Consequently, L from step 2 is a suitable Lipschitz constant. But now from (47) we get the stronger estimation

$$\|U - U^N\|_T \leq e^{LT}(\varepsilon_1 + \varepsilon_o T) =: \varepsilon \qquad \text{for } N \geq N_o ,$$

i.e. $\|V-V^N\| \leq \varepsilon_1$ implies $\|U-U^N\| \leq \varepsilon$ and therefore

$$P(\|U-U^N\|_T < \varepsilon) \leq P(\|V-V^N\|_T > \varepsilon_1) \qquad \text{for } N \geq N_o .$$

Hence for all $\varepsilon > 0$

$$\lim_{N\to\infty} P(\|U-U^N\|_T > \varepsilon) = 0$$

implying the uniform convergence in probability and the P-almost sure convergence of a suitable subsequence,

$$\lim_{R\to\infty} \|U_{tx}(\omega) - U^{N_R}_{tx}(\omega)\|_T = 0 .$$

∎

Remarks: 1) The central resultt, namely Theorem 7, states the fact, that the formal approximation (19) done in physicist's literature is very strong and can be founded by rigorous mathematics. If all Wiener processes $W_i(t)$ in (19) are constructed from the Wiener field W_{tx}, which occurs in the infinite-dimensional problem, then we arrive at the uniform convergence of the pathes in probability.

3) The estimations being necessary for the result above are of more technical nature and have been performed for the operators Δ und Δ_N. But a generalization of the statements seems to be straightforward, since we need only the asymptotic behaviour of the eigenvalues $\lambda_n \sim n^2$ and the order 1/N of the approximation of e_n by e_n^N and of λ_n by λ_n^N, respectively.

3. Some further remarks

1) From the uniform convergence of the pathes on $C([0,T]\times[0,L])$ we can immediately derive the weak convergence of the induced measures,

$$P_{U^N} \underset{N\to\infty}{\Rightarrow} P_U \quad .$$

Again this result can be extended to the space $C([0,\infty)\times[0,L])$.

2) In [12] a Girsanov formula connecting the distributions of the process U and the Gaussian process V being the solution of the linear problem was proved. Then the weak convergence of the induced measures can also be proved by using this Girsanov formula, since the well-known Girsanov formula connecting the distributions of U^N and V^N is a direct discretization of the continuous space formula [13].

3) The transition probabilities and the one-dimensional distributions of the lattice approximation (U_t^N), $t>0$, if restricted to the lattice points $x=il$, $i=1,\dots,N-1$, have densities $p_N(t,q)$ satisfying - according to (19) - the forward equation

$$\frac{\partial p_N(t,q)}{\partial t} = -\sum_{i=1}^{N-1}\frac{\partial}{\partial q_i}\left([\frac{1}{l^2}(q_{i+1}+q_{i-1}-2q_i)+f(q_i)]p_N\right) + \frac{\sigma^2}{2l}\sum_{i=1}^{N-1}\frac{\partial^2 p_N}{\partial q_i^2}$$

with $q_0=q_N=0$ and some initial distribution. But this is exactly the Fokker-Planck equation used in the physic's literature to approximate the spatially continuous system (9).

Especially for the density p_N^o of the unique invariant distribution P_N^o of the discretized system we obtain

$$p_N^o(q_1,\dots,q_{N-1}) = \tilde{N}_N\cdot\exp\left(-\frac{2l}{\sigma^2}\cdot S_N(q_1,\dots,q_{N-1})\right) ,$$

$$S_N(q) := \frac{1}{2}\sum_{i=1}^{N-1}(q_{i+1}-q_i)^2/l^2 - \sum_{i=1}^{N-1}F(q_i) , \tag{48}$$

with $F'(u)\equiv f(u)$ and $\tilde{N}_N$ as a normalization factor. This is clear due to the fact that the approximating deterministic system (19) with $\sigma=0$ can be written in gradient form,

$$\frac{du_i}{dt}(t) = -\frac{\partial S_N}{\partial u_i}[u_1(t),\dots,u_{N-1}(t)] \quad , \quad i=1,\dots,N-1 \quad .$$

In the linear case with $f\equiv 0$ P_N^o is an $(N-1)$-dimensional Gaussian distribution $Q^o:=N(0,C_N)$ with expectation zero and covariance matrix C_N, where from (48) we see

$$C_N^{-1} := -\frac{2l}{\sigma^2}\cdot\Delta_N \; , \quad \Delta_N = \frac{1}{l^2}\begin{pmatrix} -2 & 1 & 0 & 0 & \dots & 0 \\ 1 & -2 & 1 & 0 & \dots & 0 \\ 0 & 1 & -2 & 1 & \dots & 0 \\ & \dots & & & \dots & \\ 0 & \dots & & & & -2 \end{pmatrix} \quad .$$

Here Δ_N is the operator of the discretized second spatial derivative. For the inverse matrix we get

$$C_N = -\frac{\sigma^2}{2l}\cdot\Delta_N^{-1} \; , \quad (C_N)_{ik} = \frac{\sigma^2 l}{2}\cdot(\min(i,k) - \frac{i\cdot k}{N}) \; , \quad i,k=1,\dots,N-1. \quad (49)$$

In the physicist's literature usually the formal limit $N\to\infty$, i.e. $l=L/N \to 0$, is performed. Then we obtain for the invariant distribution P^o of the limiting process U a formal density which is exactly given by (10) since we have $\quad S(q) = \lim_{N\to\infty} S_N(q) \quad .$

Especially for $f\equiv 0$ this is (taking a willing interpretation) a Gaussian measure $Q^o=N(0,C)$ with covariance operator

$$C = \lim_{N\to\infty} C_N = (\sigma^2/2)\cdot(-\Delta)^{-1}$$

and this is exactly the invariant distribution of the linear equation! If $(C_N)_{ik}$ in equation (37) is interpreted as the expectation EU_iU_k with respect to P^o we obtain in the limit $N\to\infty$ for $i,k\to\infty$, $il\to x$, $kl\to y$

$$EU(x)U(y) = \frac{\sigma^2}{2}\cdot(\min(x,y) - \frac{xy}{L}) \quad , \quad x,y\in[0,L] \quad ,$$

which reproduces precisely the correlations of the spatially continuous Gaussian field $U_t(x)\equiv V_t(x)$ for given stationary distribution $P^o\equiv Q^o$, since Q^o is the distribution of a Brownian bridge on $[0,L]$.

All these results show that good physical intuition can be a very strong tool to obtain new mathematical results or at least suggestions.

4) The lattice approximation can be used to prove a result on Simulated Annealing: Suppose that the noise intensity σ decreases slowly in time, $\sigma^2(t) > \text{const}/(\log t)$. Then the one-dimensional distributions P_{U_t} converge weakly as $t\to\infty$ to a suitably weighted distribution concentrated on the global minima of the functional S.

References

[1] FARIS,W.G.,JONA-LASINIO,G.: Large fluctuations for a nonlinear heat equation with noise, J. Phys. A 15 (1982), 3025-3055

[2] FEISTEL,R.,EBELING,W.,MALCHOW,H.,SCHIMANSKY-GEIER,L.: Stochastic bifurcations in a bistable reaction-diffusion system with Neumann boundary conditions, Ann.d.Phys. 40 (1983), 151-160

[3] FIFE, P.C.: Mathematical aspects of reacting and diffusing systems, Lect. Notes Biomath., vol. 28, Springer, Berlin/Heidelberg 1979

[4] FRIEDMAN,A.: Stochastic Differential Equations and Applications, vol. I, Academic Press, New York 1975

[5] FUNAKI,T.: Random motion of strings and related evolution equations, Nagoya Math. J. 89 (1983), 129-193

[6] HAKEN, H.: Synergetics. An Introduction, 3rd ed., Springer, Berlin/Heidelberg 1983

[7] HOHENBERG, P.C., HALPERIN, B.I.: Theory of dynamical critical phenomena, Rev. Mod. Phys. 49 (1977), 435-479

[8] IWATA, K.: An infinite dimensional stochastic differential equation with state space C(ℝ), Prob. Theory Rel. Fields 74 (1987), 141-159

[9] JETSCHKE, G.: General stability analysis of dissipative structures in reaction diffusion equations with one degree of freedom, Phys. Lett. 72A (1979), 265-268

[10] JETSCHKE, G.: On the equivalence of different approaches to stochastic partial differential equations, Math. Nachr. 128 (1986), 315-329

[11] JETSCHKE, G.: Most probable states and pathes of a nonlinear reaction-diffusion equation with white noise, subm. J. Phys. A , (Preprints FSU Jena N/86/11, N/86/20, N/86/42)

[12] JETSCHKE, G., STIEWE, CH.: Transition probability and invariant distribution of a nonlinear stochastic partial differential equation, Fo.-Erg. FSU Jena N/89/23

[13] JETSCHKE, G.: Unendlichdimensionale Diffusionsprozesse als Lösungen stochastischer partieller Differentialgleichungen, Master Thesis, Jena 1989

[14] JONA-LASINIO, G., MITTER, P.K.: On the stochastic quantization of field theory, Comm. Math. Phys. 101 (1985), 409-436

[15] MA, S.: Modern theory of critical phenomena, Benjamin, New York 1976

[16] MANTHEY, R.: Existence and uniqueness of a solution of a reaction-diffusion equation with polynomial nonlinearity and white noise disturbance, Math. Nachr. 125 (1986), 121-133

[17] McKEAN, H.: Stochastic Integrals, Academic Press, London 1969

[18] METIU,M., KITAHARA,K., ROSS,J.: Stochastic theory of the kinetics of phase transitions, J. Chem. Phys. 64 (1976), 292-299

[19] PATASHINSKII, A.S., POKROVSKII, V.L.: Fluctuation theory of phase transitions, Pergamon, Oxford 1979

[20] STROOCK, D.W., VARADHAN, S.R.S.: Multidimensional diffusion processes, Springer, Berlin/Heidelberg 1979

[21] GEMAN, S., CHII-RUEY, H.: Diffusions for global optimization, SIAM J. Contr. Opt. 24 (1986), 1031-1043

LARGE TIME BEHAVIOUR OF MOMENTS OF FUNDAMENTAL SOLUTIONS OF THE RANDOM PARABOLIC EQUATION.

W. Kirsch [1] and L.A. Pastur [2]

1 INTRODUCTION

Let $V(x)$, $x \in \mathbb{R}^d$, be a random ergodic field in $\mathbb{R}^d$ (potential) and $P_\pm(t,x,y)$, $t \geq 0$, $x, y \in \mathbb{R}^d$ be the fundamental solution of the equation

$$\frac{\partial P_\pm}{\partial t} = \Delta P_\pm \pm V(x) P_\pm, \quad P|_{t=0} = \delta(x-y) \tag{1.1}$$

$$= -H^\pm P_\pm \tag{1.2}$$

where

$$H^\pm = -\Delta \pm V \tag{1.3}$$

is the Schrödinger opertator in $L^2(\mathbb{R})$ with a random potential $V(x)$. Denote by $E\{...\}$ the mathematical expectation corresponding to V. The simplest probabilistic characteristic of the random functions $P_\pm(t,x,y)$ is

$$E\{P_\pm(t,0,0)\} \equiv \tilde{N}(t) \tag{1.4}$$

According to [1], if

$$E\{e^{\pm tV(0)}\} < \infty, \quad \forall t > 0 \tag{1.5}$$

then expectation (1.4) exists and

$$\tilde{N}_\pm(t) = \int_{-\infty}^{\infty} e^{-\lambda t} N_\pm(d\lambda)$$

---

[1] Institut für Mathematik, Ruhr - Universität Bochum, D-4630 Bochum, West-Germany and SFB 237 : Unordnung und große Fluktuationen

[2] Institute for Low Temperature Physics and Engeneering, Academy of Sciences of the Ukrainian SSR, 310164 Kh'arkov, USSR

where $N_\pm(d\lambda)$ is the measure on $\mathbb{R}$, whose distribution function $N_\pm(\lambda) = N_\pm((-\infty, \lambda])$ is called integrated density of states (IDS) of the operators (1.3). The IDS plays an important role in the spectral theory of the Schrödinger opertor with a random or almost periodic potential and the theory of disordered systems (see e.g. [2] and [3] resp.). IDS can be definded as follows. Consider a family of finitely growing cubes $\Lambda \subset \mathbb{R}^d$ with the centers in the origin and sides parallel to the coordinate axes. Let H_Λ be the selfadjoint Schrödinger operator in $L^2(\Lambda)$ with boundary conditions of the form

$$\left[(1 - a(x))\frac{\partial\psi}{\partial n} + a\psi\right] |_{\partial\Lambda} = 0, \quad 0 \le a(x) \le 1, \; x \in \partial\Lambda \tag{1.6}$$

and $\{\Lambda_m^\pm\}$ be the eigenvalue of H_Λ. Set

$$N_\Lambda(d\lambda) = |\Lambda|^{-1} \sum_{\lambda_m^\pm \in d\lambda} 1 \tag{1.7}$$

where $|\Lambda|$ is the volume of the cube Λ. Then, according to [1], if potential $V(x)$ satisfies (1.5), there exists a nonrandom measure $N(d\lambda)$ (IDS), which is independent of $a(x)$ from (1.6), such that with probability 1 in the weak sense :

$$\lim_{\Lambda \to \mathbb{R}^d} N_\Lambda(d\lambda) = N(\lambda) \tag{1.8}$$

Thus, the IDS is the thermodynamic limit of (1.7). The large - time behaviour of $\tilde{N}_\pm(t)$ allows us to determine the behaviour of $N_\pm(\lambda)$ in the vicinity of the lower bottom of the spectrum of $H_\pm$. For reviews of respective results see e.g. [1-4].

Higher moments of $P(t, x, y)$ are studied in the intermittency theory (see e.g. [5]).

In this paper we study the large time behaviour of the quantity

$$\tilde{A}(t) = \int_{\mathbb{R}^d} dx E\{P_+(t, 0, x) P_-(t, x, 0)\} \tag{1.9}$$

Like $\tilde{N}(t)$ this quantity is the Laplace transform of a measure $A(d\lambda)$ which is the thermodynamic limit of random measures.

$$A_\Lambda(d\lambda) = |\Lambda|^{-1} \sum_{\lambda_n^- + \lambda_m^+ \in d\lambda} |(\psi_n^-, \psi_m^+)|^2$$

where $\psi_m^\pm$ are the eigenfunctions of $H_\Lambda^\pm$. More exactly, we have

Theorem 1 : Let a random ergodic potential satisfy the condition

$$E\{e^{tV(0)}\} < \infty \ , \ \forall t \in \mathbb{R}$$

Then there exists a nonrandom measure $A(d\lambda)$ which is independent on the boundary condition (1.6) such that with probability one in the weak sense

$$\lim_{\Lambda \uparrow \mathbb{R}^d} A_\Lambda(d\lambda) = A(d\lambda)$$

and

$$\tilde{A}(t) = \int e^{-\lambda t} A(d\lambda)$$

where $\tilde{A}(t)$ is given by (1.9).

The density of $A(\lambda)$ is (up to a certain factor) the interband light absorption coefficient of a doped semiconductor [6].

2 THE GAUSSIAN RANDOM POTENTIAL.

By using the Feynman-Kac formula for $P_\pm(t,x,y)$ we have from (1.9),

$$\tilde{A}(t) = \int_{\mathbb{R}^d} dx E\left\{ W_{(0,0)}^{(t,x)} \left\{ e^{-\int_0^t V(x(s))\, ds} \right\} W_{(0,x)}^{(t,0)} \left\{ e^{\int_0^t V(y(s))\, ds} \right\} \right\} p^2(t,x) \tag{2.1}$$

where $p(t,x) = (4\pi t)^{-d/2} e^{-x^2/4t}$ and $W_{(0,x)}^{(t,y)}\{...\}$ is the Wiener integral over trajectories $x(s), 0 \le s < t, x(0) = x, x(t) = y)$

If $V(x)$ is a Gaussian field and $E\{V(x)\} = 0$, $E\{V(0)V(x)\} = B(x)$, then

$$\tilde{A}(t) = \int_{\mathbb{R}^d} dx\, p^2(t,x)\, W_{(0,0)}^{(t,x)} \cdot W_{(0,x)}^{(t,0)} \left\{ e^{S[x(\cdot),y(\cdot)]} \right\} \tag{2.2}$$

where

$$S(x(\cdot), y(\cdot)) = \int_0^t ds_1 \int_0^t ds_2 \left[B(x(s_1) - x(s_2)) + B(y(s_1) - y(s_2)) - 2B(x(s_1) - y(s_2)) \right]$$

it is easy to show that $S \le t^2[B(0) - B(x_0)]$, where $B(x_0) = \inf_{x \in \mathbb{R}^d} B(x) \le 0$ (if $B(x) \ge 0$, then we set $x_0 = \infty$). Thus

$$\tilde{A}(t) \le (8\pi t)^{-d/2} \exp \frac{t^2}{2} [B(0) - B(x_0)] \tag{2.3}$$

The right hand side of this formula coincides up to the factor $(8\pi t)^{-d/2}$ is $e^{s}|_{x(s)\equiv 0, y(s)\equiv x_0}$. Thus, to obtain the lower bound of the same form, it is natural to restrict the integration over x to a small neighbourhood of x_0 and the Wiener integration in (2.2) to the set $\Omega_1 \times \Omega_2$, $\Omega_1 = \{|x(s)| \leq \epsilon,\ 0 \leq s \leq t-\delta\}$, $\Omega_2 = \{|y(s) - x_0| < \epsilon,\ 0 \leq s \leq t - \delta\}$, where ϵ and δ are small enough. The Wiener probability of the set Ω_1, and Ω_2 behave as $C_1 e^{-C_2 t/\epsilon^2}$, $C_1, C_2 > 0$ as $t \to \infty$. and the functional S from (2.2) is not less then $\frac{t^2}{2}[B(0) - B(x_0)](1 + o(1))$, $t \to \infty$, $\epsilon, \delta \downarrow 0$. Combining these arguments and (2.3) we obtain

Theorem 2 : Let $V(x)$ be the Gaussian random field in $\mathbb{R}^d$ with the mean zero and a correlation function $B(x)$. Then for the quantity $\tilde{A}(t)$ of (1.9) we have

$$\lim_{t\to\infty} t^{-2} \ln \tilde{A}(t) = B(0) - B(x_0) \tag{2.4}$$

where $B(x_0) = \inf_{x\in\mathbb{R}^d} B(x)$.

3 OTHER RANDOM POTENTIALS.

We discribe shortly the results for two other random potential, so called Poisson and alloy - type ones, that are widely used in the theory of disordered systems. (see, e.g. [4]). These results are obtained by using the same strategy as for the Gaussian potential but respective techniques are more complicated.

Theorem 3 : Let $f(x) \in L^1(\mathbb{R}^d) \cap C^2(\mathbb{R}^d)$, $\sup_{x\in R^d} f(x) = f(0)$; $\inf_{x\in\mathbb{R}^d} f(x) = f(x_0) \leq 0$ and $\{x_j\}$ is the Poisson point field in $\mathbb{R}^d$ with density N. Set

$$V(x) = \sum_j f(x - x_j) \tag{3.1}$$

Then

$$\ln \tilde{A}(t) = n(\frac{2\pi}{t})^{d/2} H^{-\frac{1}{2}} e^{t[f(0)-f(x_0)]}(1 + o(1)\ ,\ t \to \infty \tag{3.2}$$

where H is the Hessian of the function $f(x) - f(x + x_0)$ at $x = 0$.

Theorem 4 : Let $f(x)$ be as in Theorem 3 above, supp$f \subset \{x \in \mathbb{R}^d, |x_i| \leq 1/2, i = 1, ..., d\}$ and $\xi_n, n \in \mathbb{Z}^d$, are i.i.d. random variables with probability variables with a probability distribution $F(d\xi)$ of non compact support. Set

$$V(x) = \sum_{n \in \mathbb{Z}^d} \xi_n f(x-n) \tag{3.3}$$

Then

$$\ln \tilde{A}(t) = \ln \left\{ \tilde{F}(tf(0) + \tilde{F}(tf(x_0) \right\} (1 + o(1)) \quad t \to \infty \tag{3.4}$$

where $\tilde{F}$ is the Laplace transform of F.

4 DISCUSSION.

All upper bounds, that we used above, obtaining asymptotic formulas (2.4), (3.2), (3.4) and a number of others, are special cases of the following general bound

$$\tilde{A}(t) \leq \int_{\mathbb{R}^d} dx p^2(t,x) E\{e^{tV(0) - tV(x)}\}$$

This bound results from the representation (21) and the Jensen inequality.

It is also possible to obtain a general lower bound, that is asymptotically exact for $t \to \infty$ in many cases (including those given in Theorems 2 - 4). To formulate this bound it is convenient to rewrite representation (2.1) in the form

$$\tilde{A}(t) = p(t,0) W^{(2t,0)}_{(0,0)} E \left\{ \exp \left[- \int_0^{2t} v_t(s, x(s))\, ds \right] \right\}$$

where

$$V_t(s,x) = \begin{cases} V(x) \ , & 0 \leq s < t \\ -V(x) \ , & t \leq s < 2t \end{cases} \tag{4.1}$$

Thus our moment (1.9) is a double expectation containing time-dependent random potential (4.1). Consider now an abitrary time-dependent nonrandom potential $V(s,x)$ and let $Q(0,s;x,y)$ be the fundamental solution of the heat equation with this potential. Then for arbitrary nonnegtive $\psi \in L^2(\mathbb{R}^d), \|\psi\|_{L^2(\mathbb{R}^d)} = 1$

$$\tilde{A}(t) \geq Z_t \|\psi\|_{L^1(\mathbb{R}^d)} E \left\{ \exp \left\{ Z_t^{-1} \int_0^{2t} ds (\psi, Q(0,s)(V-U) \cdot Q(s,t)\psi) \right\} \right\}$$

where $Z_t = (\psi, Q(0,2t)\psi)$ and $Q(s,t)$ is the operator, defined in $L^2(\mathbb{R}^D)$ by the kernel $Q(s,t;x,y)$. The derivation of this bound and a variety of its application will be given elsewhere.

REFERENCES

1. L.A. Pastur. Teor.Mat.Fiz. 32 (1977) 88-95 (Russian)

2. L.A. Pastur. Soviet Sci. Rev. 60 (1986) 1-112

3. I.M.Lifshitz, S.A.Gredeskul, L.A.Pastur. Introduction to the Theorie of disordered Systems. Whiley, New York, 1988

4. W.Kirsch. Lecture Notes in Mathem. 1109 (1985) 68-102

5. Ya.B.Zel'dovich, S.A.Molchanov, A.A.Ruzmaikin, D.D.Sololov. Usp. Fiz. Nauk, 152 (1986) 3-32

6. A.L.Efros, B.L. Shklovski. Electronic Properties of Doped Semiconductors. Berlin, Springer, 1984

A Comparison Theorem for a Class of Stochastic Partial Differential Equations

Peter Kotelenez
Department of Mathematics & Statistics
and Center for Stochastic and Chaotic Processes in Science and Technology
Case Western Reserve University
Cleveland, Ohio 44106, USA

Abstract

A comparison theorem is derived for a class of stochastic partial differential equations (SPDE's). The SPDE's are defined on a function space, are driven by a function valued (possibly cylindrical) Wiener process, and the coefficients are time dependent. This generalizes a result proved in Kotelenez [2]. As a corollary a sufficient condition for positivity of solutions is obtained.

Let $(\Omega, \mathcal{F}, \mathcal{F}_{t\geq 0}, P)$ be a stochastic basis with right continuous filtration. Further, let $O \subset \mathbf{R}^n$ be some domain and denote by r a generic element of $O.\mu$ is a measure on the Borel σ-algebra of O and $\mathbf{H} := L_2(O, d\mu)$, the space of real valued $d\mu$-square integrable functions on $O.|\cdot|$ and $\langle \cdot, \cdot \rangle$ will denote the norm and the scalar product on $\mathbf{H}$, respectively. $A(t)$ is a family of possibly unbounded closed operators on $\mathbf{H}$ which generates an evolution operator $U(t,s)$ (i.e., a two-parameter strongly continuous semigroup - cf. Curtain and Pritchard [1] as well as Tanabe [3], where $U(t,s)$ is called a fundamental solution). R^i and $F, i = 1,2$ are functions from $\mathbf{R}_+ \times \mathbf{R}$ into $\mathbf{R}$, and $W(t)$ is an $\mathbf{H}$-valued (possibly cylindrical) Wiener process. Let us consider the following stochastic evolution equation on $\mathbf{H}$:

$$dX^i = [A(t)X^i + R^i(t, X^i)]dt + F(t, X^i)dW, \quad X^i(0) = X_0^i \in \mathbf{H}_i = 1,2 \quad (1)$$

where $F(t,X) = F(t, X(\cdot))$ $(X(\cdot) \in \mathbf{H})$ acts as a multiplication operator on $\mathbf{H}$. We want to derive a comparison theorem for the mild solutions of (1), i.e., for the solutions of the integral equations:

$$X^i(t) = U(t,0)X_0^i + \int_0^t U(t,s)[R^i(s, X^i(s))ds + F(s, X^i(s))]dW(s)$$
$$i = 1,2. \quad (2)$$

Following the ideas in [2] we decompose (1) into its deterministic and into its fluctuation parts, prove comparison theorems for the separated parts, and finally derive the comparison theorem for (1) via a suitable form of the Trotter product formula. Consequently, consider

$$\begin{aligned} dY^i &= [A(t)Y^i + R^i(t,Y^i)]dt \\ Y^i(0) &= Y_0^i \in \mathbf{H}, i = 1,2, \end{aligned} \tag{3}$$

and

$$\begin{aligned} dZ^i &= F(t,Z^i)dW \\ Z^i(0) &= Z_o^i \in \mathbf{H}. \end{aligned} \tag{4}$$

Note that (4) may not have a solution, if $W(t)$ is cylindrical, whereas (1), resp. (2), can have a solution if $U(t,s)$ is sufficiently smoothening. We will circumvent this problem by assuming that (4) can be approximately solved. Let us first investigate (3) and assume.

A.1

(i) $R^1(t,x) \le R^2(t,x)$ for all $(t,x) \in \mathbf{R}_+ \times \mathbf{R}$;

(ii) there are sequences of functions $R_L^i, i = 1,2, L \in \mathbf{N}$, such that $|R_L^i(t,x) - R_L^i(t,y)| \le K_L|x-y|, K_L < \infty$ for all $t \ge 0, x,y \in \mathbf{R}$ and $|R_L^i(t,0)| \le K_{T,L} < \infty$ for all $t \le T$. Moreover, for all $t \ge 0, x \in \mathbf{R}$

$$R^i(t,x) = \lim_{L\to\infty} R_L^i(t,x) :$$

(iii) (3) has unique mild solutions $Y_i, i = 1,2$, and, denoting by Y_L^i the mild solutions of (3) if R^i is replaced by R_L^i,

$$|Y^i(t) - Y_L^i(t)| \to 0, \text{as } L \to \infty \text{ for all } t \ge 0.$$

(iv) $U(t,s)$ is positivity preserving.

If $f,g \in \mathbf{H}$ we will write $f \ge g$ if $f(r) \ge g(r)d\mu$ almost everywhere.

Lemma 1

Suppose $Y_0^1 \le Y_0^2$ in addition to A.1. Then for any $t \ge 0$

$$Y^1(t) \le Y^2(t) \tag{5}$$

Proof

We may without loss of generality assume that the R^i's have the properties of the R^i_L's given in A.1 (ii). Hence

$$A(s)+\frac{R^2(s,Y^2(s))-R^2(s,Y^1(s))}{Y^2(s)-Y^1(s)}$$

generates a positivity preserving evolution operator $V(t,s)$, where the fraction (defined pointwise in $r \in O$ and set to equal to 0 if $Y^2(s,r)=Y^1(s,r)$) has to be interpreted as a multiplication operator (cf. Kotelenez [2]).

Thus,

$$Y^2(t)-Y^1(t)=V(t,0)(Y_0^2-Y_0^1)+\int_0^t V(t,s)[R^2(s,Y^1(s))-R^1(s,Y^1(s))]ds \geq 0$$

for all $t \geq 0$. ∎

We are now considering the fluctuation part of (1) and make the following assumptions (always supposing non-anticipating **H**-valued initial conditions):

A.2.

(i) There is a complete orthonormal system of continuous functions $\{\phi_n\}$ for **H**.

(ii) Setting $W_M(t,r) := \sum_{i=1}^M < W(t),\phi_i > \phi_i(r)$,

$$dZ^M(t,r)=F(t,Z^M(t,r))dW_M(t,r) \tag{6}$$

is uniquely solvable as an Itô-equation both for any $r \in O$ and as an **H**-valued equation.

(iii) If $Z_0^k \to Z_0$ in **H**, then $Z_m^k(t,Z_0^k) \to Z_M(t,Z_0)$ as $k\to\infty$, where $Z^k(t,Z_0^k)$ and $Z(t,Z_0)$ are the **H**-valued solutions of (6) with initial conditions Z_0^k and Z_0, respectively.

(iv)

$$dX_M^i=[A(t)X_M^i+R^i(t,X_M^i)]dt+F(t,X_M^i)dW_m \tag{7}$$

has a unique mild (Itô-) solution in $\mathbf{H}, i=1,2$.

(v) Denoting by λ_T the Lebesque measure on $[O,T]$ and by $\lambda_T\otimes\mu\otimes P$ the product measure we suppose that for $i=1,2$ $X_M^i \to X^i$ $\lambda_T\otimes\mu\otimes P$-stochastically for any $T>0$ if $M\to\infty$ and $X_M^i(0)=X^i(0)$.

The following lemma is obvious in view of A.2, (ii) and (iii), (in particular the uniqueness of the associated one-dimensional Itô-equations):

Lemma 2
Suppose $Z_0^1 \leq Z_0^2$. Then $Z_M^1(t, Z_0^1) \leq Z_M^2(t, Z_0^2)$.

Now we can prove the main result. We denote by $X^i(t, X_0^i)$ the mild solutions of (1) with initial condition X_0^i.

Theorem (Comparison Theorem)
Assume $X_0^1 \leq X_0^2$ on $\mathbf{H}, X_0^i$ $\mathcal{F}_0$-adapted, and $E|X_0^i|^2 < \infty, i = 1,2$. Assume that (1) has unique mild solutions for $i = 1,2$, and $t \leq T < \infty$ and suppose A.1 and A.2. Then as for all $t \in [0,T]$

$$X^1(t, X_0^1) \leq X^2(t, X_0^2). \tag{8}$$

Proof

(i) We make a state space extension to obtain a time-homogeneous Markov process:

$$\widetilde{X} := (t, X) \in [0,T] \times \mathbf{H} =: \widetilde{H}$$

Thus, equations (1) can be rewritten as

$$d\,\widetilde{X}^{\,i} = \widetilde{A}^{\,i}(\widetilde{X}^{\,i})dt + \widetilde{F}\,(\widetilde{X}^{\,i})dW, \tag{9}$$

with $\widetilde{A}^{\,i}(\widetilde{X}) = (1, A(t)X + R^i(t,X)), i = 1,2$, and $\widetilde{F}\,(\widetilde{X}) = (0, F(t,X))$. Equations (3) and (4) can be written in the same fashion. Obviously $\widetilde{Y}^{\,1}(t, y_0^1) \leq \widetilde{Y}^{\,2}(t, y_0^2)$ if $Y_0^1 \leq Y_0^2$ (by Lemma 1, where $(s,X) =: \widetilde{X} \leq \widetilde{Y} = (t,Y)$ if and only if $s \leq t$ and $X \leq Y$). Similarly, $\widetilde{Z}\,(t, Z_0^1) \leq \widetilde{Z}\,(t, Z_0^2)$ if $Z_0^1 \leq Z_0^2$.

(ii) Let $\Gamma = \{\widetilde{X}, \widetilde{Y}) \in \widetilde{\mathbf{H}} \times \widetilde{\mathbf{H}} : \widetilde{X} = (t,X)$ and $\widetilde{Y} = (t,Y)$ and $X \leq Y\}$. Since both $(\widetilde{Y}^{1}(t), \widetilde{Y}^{2}(t))$ and $(\widetilde{Z}^{\,1}(t), \widetilde{Z}^{\,2}(t))$ are Markov processes on $\widetilde{\mathbf{H}} \times \widetilde{\mathbf{H}}$ we may apply the Trotter product formula to their associates semigroups $\mathbf{P}_{\widetilde{Y}}(t)$ and $\mathbf{P}_{\widetilde{Z}}(t)$ on $C_b(\widetilde{\mathbf{H}} \times \widetilde{\mathbf{H}}, \mathbf{R})$, the space of bounded continuous functions from $\widetilde{\mathbf{H}} \times \widetilde{\mathbf{H}}$ into $\mathbf{R}$. Following the procedure in Kotelenez [2] we obtain that the Markov semigroup $\mathbf{P}_{\widetilde{X}}(t)$ generated by $(\widetilde{X}^{1}(t), \widetilde{X}^{2}(t))$ is given by

$$\mathbf{P}_{\widetilde{X}}(t)f = \lim_{n \to \infty}(\mathbf{P}_{\widetilde{Y}}(\frac{t}{n})\mathbf{P}_{\widetilde{Z}}(\frac{t}{n}))^n f \tag{10}$$

for all f in a core of the generator of $\mathbf{P}_{\widetilde{X}}(t)$. Now the comparison theorems for Y^i and Z^i imply that $(\mathbf{P}_{\widetilde{Y}} 1_\Gamma)(\xi) \equiv 1 \equiv (\mathbf{P}_{\widetilde{Z}}(t)1_\Gamma)(\xi)$ for $t \in [0,T]$ if $\xi \in \Gamma$, where $1_{\widetilde{\Gamma}}(\xi) = 1$ if $\xi \in \Gamma$ and $= 0$ otherwise. By approximating 1_Γ by a sequence f_n, satisfying (10) (cf. [2]) we obtain

$$(\mathbf{P}_{\widetilde{X}}(t)1_\Gamma)(\xi) \equiv 1, \ if\ \xi \in \Gamma, \tag{11}$$

which implies (8). ∎

Next we consider the case of only one equation of type (1), i.e., for only one function $R(t,x)$.

Corollary (Positivity)

In addition to A.1 and A.2 suppose for any $\mathcal{F}_0$-measurable X_0 such that $E|X_0|^2 < \infty$ (1) has a unique mild solution $X(\cdot, X_0)$ on $[0,T]$ and $R(t,0) \geq 0$ for all $t \in [0,T]$. Moreover, suppose $F(t,0) \equiv 0$ for all $t \in [0,T]$. Then

$$X_0 \geq 0 \; implies \; X(t,X_0) \geq 0 \; for\, all\, t \in [0,T]. \tag{12}$$

Proof

Setting $R^2(t,x) := R(t,x), R^1(t,x) := R(t,x) - R(t,0), X_0^2 := X_0 \geq 0$ and $X_0^1 \equiv 0$ we obtain from the comparison theorem and our assumptions:

$$X(t,X_0) =: X^2(t,X_0^2) \geq X^1(t,X_0^1) \equiv 0.$$

∎

Remark

In the time-homogeneous case it was shown in [2] that $A \sim T(t)$ positivity preserving, $R(0) \geq 0$ and $F(0) = 0$ are also necessary for the positivity of the mild solution of (1). Moreover, existence, uniqueness and smoothness was proved for the mild solution of (1) for $W(t)$ the $L_2(O, d\mu)$ valued standard cylindrical process, R and F Lipschitz, and $A(t)$ from a wide class of pseudo-differential operators ($d\mu$ being the Lebesque measure multiplied with a certain weight function.) For positive solutions of (1) with time-homogeneous coefficients, the Lipschitz condition on R was relaxed to include a class of polynomials.

REFERENCES

1. Curtain, R.F. and Pritchard, A.J., Infinite dimensional linear systems theory: LN in Control and Information Sciences, Vol. 8. Berlin-New York: Springer 1978.

2. Kotelenez, P.: A class of function and density valued stochastic partial differential equations driven by space-time white noise. Preprint No. 89-173, Case Western Reserve University, Department of Mathematics and Statistics (1989).

3. Tanabe, M.: Equations of evolution. London: Pitman 1979.

WEAK APPROXIMATION OF A NONLINEAR STOCHASTIC PARTIAL DIFFERENTIAL EQUATION

Ralf Manthey
Friedrich-Schiller-Universität
Mathematische Fakultät, UHH 17.OG
J e n a, 0-6900, Germany

1. Introduction

Consider the following formal initial-boundary value problem

$$(D)\quad \begin{cases} \frac{\partial}{\partial t}u(t,x) = \frac{\partial^2}{\partial x^2}u(t,x) + f(u(t,x)) + \sigma\xi(t,x), & t>0, x\in(0,1), \\ u(0,x) = u_0(x), & x\in[0,1], \\ u(t,0) = u(t,1) = 0, & t\geq 0, \end{cases}$$

which was investigated by many authors from quite different viewpoints (cf. [1],[3],[6]-[10]). Let $(\Omega,\mathfrak{F},\mathbb{P})$ denote the basic complete probability space and suppose that $u_0:\Omega\times[0,1]\to\mathbb{R}$ is pathwise continuous. If ξ is a space-time GAUSSian white noise, i.e. a centered GAUSSian $\mathcal{D}'(\mathbb{R}_+\times[0,1])$-valued random element with covariance functional

$$\mathbb{E}\xi(\varphi)\xi(\psi) = \int_{\mathbb{R}_+}\int_0^1 \varphi(t,x)\psi(t,x)\,dx\,dt,$$

then (D) represents only a symbol ($\mathcal{D}'$ denotes the space of SCHWARTZ distributions.). Indeed, in the considered case of a one-dimensional space parameter $x\in[0,1]$ it is possible to give (D) a precise mathematical meaning as follows. Let W denote a BROWNian sheet, i.e. a pathwise continuous centered GAUSSian random field defined on $\mathbb{R}^2_+$ with covariance function

$$\mathbb{E}W(t,x)W(s,y) = (t\wedge s)(x\wedge y).$$

Introduce the σ-algebra $\mathfrak{F}_t = \sigma\{u_0(x), W(s,x):\ 0\leq s\leq t,\ x\in[0,1]\}$.

1.1 Definition. *A random field $u=(u(t,x))_{t\geq 0, x\in[0,1]}$ is called a solution to (D) if it is pathwise continuous and $\mathfrak{F}_t$-measurable for any $t\geq 0$ and satisfies the stochastic integral equation*

$$
\text{(I)}\quad
\begin{aligned}
u(t,x) &= \int_0^1 G(t,x,y)u_0(y)\,dy + \int_0^t\int_0^1 G(t-s,x,y)f(u(s,y))\,dy\,ds \\
&\quad + \sigma\cdot\int_0^t\int_0^1 G(t-s,x,y)\,dW_{sy} \\
&= a(t,x) + F(u)(t,x) + \sigma\cdot r(t,x),
\end{aligned}
$$

$\mathbb{P}$ *a.s. for any* $(t,x)\in[0,\infty)\times[0,1]$.

Here G is the GREEN's function of the corresponding linear problem (D), i.e.

$$G(t,x,y) = \sum_{k=1}^{\infty} h_k(x)h_k(y)\exp(-\lambda_k t),$$

where $h_k(x)=\sqrt{2}\cdot\sin\pi kx$, $k=1,2,\ldots,$ are the eigenfunctions of the negative Laplace operator with boundary conditions as considered above. Each h_k corresponds to the eigenvalue $\lambda_k=(k\pi)^2$.

One of the arguments to consider the model with a space-time white noise is the fact that in physically interesting cases the values of the real noise are essential weaker correlated as those of the state of the considered system. Moreover, the use of the white noise leads to the MARKOV property of the solution and to a stationary distribution which is well-known in physics, see for example [12].

Nevertheless, it makes sense to consider (D) with correlated noises. In this case there are no difficulties to define the meaning of (D) in any dimensions of the space parameter x.

In this paper the natural question under which conditions the solutions of (D) with a correlated noise weakly converge to a solution of (D) with a white noise disturbance get an answer. The case of a physical interesting (polynomial) nonlinearity is included as well as the standard LIPSCHITZ case.

On the other side one can construct approximations of the solution to (D) by considering a "point noise" instead of the GAUSSian white noise. How to do this will be shown in the last part of the paper.

2. The linear case ($f=0, u_0=0$)

Fix $T>0$ and consider $(t,x)\in\mathbb{D}_T=[0,T]\times[0,1]$. In the linear case the solution of (D) is defined to be $r=(r(t,x))_{(t,x)\in\mathbb{D}_T}$. The centered GAUSSian random field r is characterized by its covariance function

$$\mathbb{E}r(t,x)r(s,y) = \int_0^{t\wedge s} G(t+s-2\tau,x,y)\,d\tau .$$

There exists a continuous version (cf. [13, [8]) which will be always considered in the sequel.
Let $(\xi_n)_{n\geq 1}$ be a sequence of centered GAUSSian random fields defined on $\Omega\times\mathbb{D}_T$. The following is assumed.

(V0) $\iint_{\mathbb{D}_T} \xi_n^2(t,x)\,dtdx < \infty$ *for any* $\omega\in\Omega$.

(V1) *For each* $m=1,2,\ldots$ *the random vector*

$$\left(\int_0^{t_1}\int_0^{x_1} \xi_n(s,y)\,dyds,\ldots,\int_0^{t_m}\int_0^{x_m} \xi_n(s,y)\,dyds\right)$$

weakly converges to $(W_{t_1,x_1},\ldots,W_{t_m,x_m})$ *for arbitrary* $(t_1,x_1),\ldots,(t_m,x_m)\in\mathbb{D}_T$.

(V2) $\limsup_{n\to\infty} \iint_{\mathbb{D}_T} |\mathbb{E}\xi_n(s,y)\xi_n(t,x)|\,dyds \leq C < \infty$ *for all* $(t,x)\in\mathbb{D}_T$.

Let

$$r_n = \left(\int_0^t\int_0^x G(t-s,x,y)\xi_n(s,y)\,dyds\right)_{(t,x)\in\mathbb{D}_T}$$

Note that $\mathbb{P}\circ r^{-1}$ and $\mathbb{P}\circ r_n^{-1}$, $n=1,2,\ldots$ are measures on the space of continuous functions defined on $\mathbb{D}_T$ denoted by $(\mathbb{C}(\mathbb{D}_T),\mathcal{B}(\mathbb{D}_T))$. Recall that a sequence of random fields converges weakly if the corresponding distribution weakly converge. This type of convergence is symbolized by "$\Rightarrow$".

THEOREM (cf.[8]). *Under the conditions* (V0) - (V2) *it holds*

$$\left(\int_0^t\int_0^x \xi_n(s,y)\,dyds\right)_{(t,x)\in\mathbb{D}_T} \Rightarrow W \quad\text{and}\quad r_n \Rightarrow r \quad \text{as } n\to\infty.$$

REMARKS.

1. If the random fields are not GAUSSian, then only the weak convergence of the finite-dimensional distributions can be shown (cf. [8]).

2. The conditions (V1) and (V2) are very natural. Formally, they are fulfilled by ξ.

EXAMPLE. Let $n=1,2,\ldots$ and consider centered GAUSSian random fields with covariance functions

$$C_n(t,x;s,y) = a_n^2 \cdot exp(-b_n(|t-s| + |x-y|)),$$

where $a_n, b_n > 0$ and $lim\ a_n^{-1} = lim\ b_n^{-1} = 0$, $lim\ a_n b_n = \frac{1}{2}$. This example of two-parameter ORNSTEIN-UHLENBECK fields was discussed in [8].

3. The nonlinear case

3.1 Existence and uniqueness of a solution

Let $\eta : \Omega \times \mathbb{D}_T \to \mathbb{R}$ be a pathwise continuous random field and consider

$$(\mathrm{I}_\eta) \qquad u(t,x) = a(t,x) + F(u)(t,x) + \eta(t,x).$$

Assume that u_o is a pathwise continuous random process. Then a has continuous trajectories.

3.1.1 THEOREM. *Let η and u_o be as above and suppose that f is globally LIPSCHITZ continuous. Then there exits a pathwise unique continuous solution to (I_η).*

The proof is routine.

3.1.2 THEOREM (cf. [9]). *Let η and u_o as above and suppose that f is locally LIPSCHITZ continuous.*

If

$$(1) \qquad \inf_{u \le 0} f(u) > -\infty$$

and

$$(2) \qquad \sup_{u \ge 0} f(u) < +\infty,$$

then there exists a pathwise unique continuous solution to (I_η).

3.2 Continuous dependence on η

THEOREM. *Under the conditions of Theorem 3.1.1 and Theorem 3.1.2 the mapping $\eta(\omega)\to u(\omega)$ from $\mathbb{C}(\mathbb{D}_T)$ into $\mathbb{C}(\mathbb{D}_T)$ is continuous for every $\omega\in\Omega$.*

P r o o f. (i) The case of Theorem 3.1.1. Introduce the notations $\|\varphi\| = \sup_{x\in[0,1]} |\varphi(x)|$ and $\|\psi\|_t = \sup_{(s,x)\in\mathbb{D}_t} |\psi(s,x)|$. Let η_1 and η_2 be two pathwise continuous random fields on $\Omega\times\mathbb{D}_T$ and consider $\eta=\eta_1-\eta_2$. For the corresponding solutions of (I_{η_i}), $i=1,2$, denoted by u_1 and u_2 let $u = u_1-u_2$. One obtains

$$|u(t,x)| \le |\eta(t,x)| + L\cdot\int_0^t\int_0^1 G(t-s,x,y)\,|u(s,y)|\,dyds$$

$$\le \|\eta\|_t + L\cdot\int_0^t \|u\|_s ds,$$

where L is the LIPSCHITZ constant. Consequently,

$$\|u\|_t \le const\cdot\|\eta\|_t$$

for any $t\in[0,T]$.

(ii) The case of Theorem 3.1.2. Note that if f is locally LIPSCHITZ continuous then (1) and (2) are equivalent to the existence of two nonincreasing functions g and h such that $g \le f \le h$. Let u and η as in the first part of the proof and assume $\|\eta\|_T<\delta$, where $\delta>0$ is a given real number. Suppose that u_1 and $\omega\in\Omega$ are fixed. Because of

$$(3)\qquad \big|\,\|a+\eta_2\|_T - \|a+\eta_1\|_T\big| \le \|\eta\|_T < \delta$$

and

$$(4)\qquad \|u_2\|_T \le \|a+\eta_2\|_T + T\cdot\Big[h(-\|a+\eta_2\|_T) - g(\|a+\eta_2\|_T)\Big]$$

there exists a random variable $N_\delta(\omega)$ which not depends on η_2 being in a δ-ball around η_1 and such that

$$\|u_2(\omega)\|_T \le N_\delta(\omega).$$

(The inequality (4) is proved in [12].) By the local LIPSCHITZ continuity and GRONWALL's Lemma one can conclude

$$\|u_1-u_2\|_T \leq exp(L_{N_r(\omega)}\cdot T)\cdot\|\eta_1-\eta_2\|_T,$$

what proves the assertion.

4. Weak convergence

Denote by u_n the solution of (I_{r_n}).

THEOREM. *Let f satisfy the conditions of Theorem 3.1.1 or Theorem 3.1.2 and suppose that* $(\xi_n)_{n=1,2,...}$ *is a family of centered GAUSSian random fields with the properties* (V0)-(V2). *Then* u_n *weakly converges to u on* $\mathbb{C}(\mathbb{D}_T)$, *where u is the unique solution to* (I).

P r o o f. Since continuous mappings preserve the weak convergence this is an immediate consequence of the theorems in Section 2 and Section 3.2.

5. Approximation by point fields

The following construction is an example of a much more general approach which will be published in the forthcoming joint paper [5]. By standard methods one can show the validity of

$$\int_0^1\left(\int_0^t\int_0^1 G(t-s,x,y)\,dW_{sy}\right)h(x)\,dx = \int_0^t\int_0^1\left(\int_0^1 G(t-s,x,y)h(x)\,dx\right)dW_{sy}$$

$\mathbb{P}$ a.s., where $h\in\mathbb{L}_2(0,1)$ (cf. [10]). Denote the r.h.s. of this equality by $\Pi_t(h)$. In order to construct an approximation for $\Pi_t(h)$ introduce the notion of a point process on the parameter space $[\mathbb{R}_+\times[0,1],\mathfrak{B},\lambda]$, where $\mathfrak{B}$ denotes the corresponding BOREL σ-algebra and λ stands for the LEBESGUE measure. Let $\mathbb{M}=\{\phi:\ \phi=\sum_{i\in I}\delta_{z_i}$, *$\phi$ is a locally finite measure on* $\mathbb{R}_+\times[0,1]\}$ and denote by $\mathfrak{M}$ the smallest σ-algebra for which the mappings $\phi\to\phi(B)$ are measurable for all $B\in\mathfrak{B}$. A mapping $\omega\to\phi(.,\omega)$ from Ω into $\mathbb{M}$ with distribution $P=\mathbb{P}\circ\phi^{-1}$ is called point process. A point process is said to be POISSONian with intensity measure λ if for any $n=2,3,...$ and BOREL sets $B_1,\ldots,B_n$ with $B_i\cap B_j=\emptyset$, $i\neq j$, the random variables

$\phi(B_1),\dots,\phi(B_n)$ are independent and $\phi(B)$ is POISSON distributed with parameter $\lambda(B)$. Let $\phi_{1,j}$, $j=1,\dots,n$, be independent POISSONian point processes with intensity measure λ and put

$$\phi_n = \sum_{j=1}^{n} \phi_{1,j}.$$

It is well-known that ϕ_n is again a POISSONian point process with intensity measure $n\cdot\lambda$. Consider a family of independent and identically distributed random variables $(\varkappa_i^{(n)})_{i\in I}$ with

$$\mathbb{P}(\{\varkappa_i^{(n)} = -n^{-1/2}\}) = \mathbb{P}(\{\varkappa_i^{(n)} = +n^{-1/2}\}) = \frac{1}{2}.$$

Put

$$\phi^{(n)} = \sum_{i\in I} \varkappa_i^{(n)}\delta_{z_i},$$

i.e. every point of ϕ_n gets independently the weight $\varkappa_i^{(n)}$. Then $\phi^{(n)}$ represents a locally finite random measure on $\mathbb{R}_+\times[0,1]$ with the properties

(ϕ1) $\quad \mathbb{E}\int h(z)\phi^{(n)}(dz) = 0,$

(ϕ2) $\quad \mathbb{E}\left[\int h(z)\phi^{(n)}(dz)\right]^2 = \int h^2(z)dz \qquad$ *for* $h\in\mathbb{L}_2(\mathbb{R}_+\times[0,1])$

and

(ϕ3) $\quad \int h_1(z)\phi^{(n)}(dz),\dots,\int h_m(z)\phi^{(n)}(dz)$ *are independent if*

$\qquad$ *supp* $h_i \cap$ *supp* $h_j = \emptyset$, $i\neq j$ *and* $h_i\in\mathbb{L}_2(\mathbb{R}_+\times[0,1])$, $i,j=1,\dots,m$

Define a mapping $W:\Omega\times\mathbb{L}_2(\mathbb{R}_+\times[0,1]) \to \mathbb{R}$ by

(W1) $\quad W(\varphi)$ *is GAUSSian with* $\mathbb{E}W(\varphi)=0$ *for any* $\varphi\in\mathbb{L}_2(\mathbb{R}_+\times[0,1])$

(W2) $\quad \mathbb{E}W(\varphi)W(\psi) = \int \varphi(z)\psi(z)\lambda(dz)$

(W3) $\quad W(\alpha_1\varphi_1 + \alpha_2\varphi_2) = \alpha_1 W(\varphi_1) + \alpha_2 W(\varphi_2) \qquad \mathbb{P}$ *a.s. for any* $\alpha_1,\alpha_2\in\mathbb{R}$, $\varphi_1,\varphi_2\in\mathbb{L}_2(\mathbb{R}_+\times[0,1])$.

If W is considered on $\mathcal{D}(\mathbb{R}_+\times[0,1])$ one obtains the GAUSSian white noise mentioned in the introduction. Furthermore, if $\varphi=1_A$, $A\in\mathfrak{B}(\mathbb{R}_+\times[0,1])$, the corresponding GAUSSian family is called set-indexed white noise with control measure λ. Note that $(W(A))_{A\in\mathfrak{B}_o}$, where $\mathfrak{B}_o=\{A\in\mathfrak{B}: \lambda(A)<\infty\}$ is not a random measure in the classical ($\mathbb{P}$ a.s.) sense but only in $\mathbb{L}_2(\Omega)$. Nevertheless, one can construct a (stochastic) integral

$$\int h(z)W(dz)$$

for any $h\in\mathbb{L}_2(\mathbb{R}_+\times[0,1])$, which possesses the same properties (ϕ1)-(ϕ3) as above if $\phi^{(n)}$ is replaced by W.
By the central limit theorem one can conclude that

$$\int h(z)\phi^{(n)}(dz) \Rightarrow W(h),$$

$h\in\mathbb{L}_2(\mathbb{R}_+\times[0,1])$, as $n\to\infty$.

Note

$$(\phi 2)^* \qquad \mathbb{E}\left(\int h(z)\phi^{(n)}(dz)\right)^m = \begin{cases} 0, & \text{if } m \text{ is odd} \\ n^{-(k-1)}\int h^{2k}(z)\,dz, & \text{if } m=2k. \end{cases}$$

Let B be a BOREL set in [0,1] and introduce the kernel

$$K_t(x,B) = \int_B G(t,x,y)\,dy.$$

Let $\varphi\in\mathbb{L}_2(0,1)$ and assume that u is a solution to (D). Consider

$$U_t*\varphi=\int_0^1 u(t,x)h(x)\,dx=(K_t*u_o)*\varphi + \int_0^t\int_0^1 (K_{t-s}*\varphi)(y)f(u(s,y))\,dyds + \Pi_t(\varphi).$$

Introduce

$$\Pi_t^{(n)}(\varphi) = \int_{\mathbb{R}_+}\int_0^1 (K_{t-s}*\varphi)(y)1_{[0,t)}(s)\phi^{(n)}(d[s,y]).$$

Since

$$\int_0^t\int_0^1\left(\int_0^1 G(t-s,x,y)\varphi(x)\,dx\right)^2 dyds$$

$$\le \int_0^t\int_0^1\left(\int_0^1 G(t-s,x,y)\varphi^2(x)\,dx\right)\left(\int_0^1 G(t-s,x,y)\,dx\right)dyds$$

$$\le t\cdot\int_0^1 \varphi^2(x)\,dx < \infty$$

the property (ϕ2) implies

$$\mathbb{E}(\Pi_t^{(n)}(\varphi))^2 = \mathbb{E}\Pi_t^2(\varphi).$$

Moreover, it holds

$$\mathbb{P}\Big[\ \Pi_t^{(n)}(\varphi) = n^{-1/2}\cdot\sum_{j=1}^{n}\int_{\mathbb{R}_+}\int_0^1 (K_{t-s}*\varphi)(y)1_{[0,t)}(s)\phi_{1,j}(d[s,y])\Big]=1$$

for any $t\in\mathbb{R}_+$, $\varphi\in\mathbb{L}_2(0,1)$.
Consequently, again by the central limit theorem

$$\Pi_t^{(n)}(\varphi)) \Rightarrow \Pi_t(\varphi)$$

for any $t\in\mathbb{R}_+$, $\varphi\in\mathbb{L}_2(0,1)$.

A similar construction was studied in [4].

LEMMA. *Let f be globally LIPSCHITZ continuous. Then the equation*

$$u_n(t,x) = (K_t*u_o)(x) + \int_0^t\int_0^1 G(t-s,x,y)f(u_n(s,y))dyds +$$

$$\int_{\mathbb{R}_+}\int_0^1 G(t-s,x,y)1_{[0,t)}(s)\phi^{(n)}(d[s,y]),$$

has a unique solution for any n=1,2,...

The main result is the following assertion.

THEOREM (cf. [5]). *Let ψ:[0,1] → $\mathbb{R}$ be a measurable and bounded function with compact support and f as in the Lemma above. Then*

$$U_t^{(n)}*\psi \Rightarrow U_t*\psi$$

for any t>0.

References

[1] G. DA PRATO, J. ZABCZYK: A Note on Semilinear Stochastic Equations, *Diff. and Integr. Equ. 1-2 (1988) 143-155*

[2] D.A. DAWSON: Stochastic Evolution Equations and Related Measure Processes, *J.Multivariate Anal. 5-1 (1975) 1-52*

[3] W.G. FARIS, G. JONA-LASINIO: Large fluctuation for a nonlinear heat equation with noise, *J.Phys.A:Math.Gen. 15 (1982) 3025-3055*

[4] K.-H. FICHTNER, M. SCHMIDT: Approximation of a continuous system by point systems, *SERDICA 13 (1987) 396-402*

[5] K.-H. FICHTNER, R. MANTHEY: Weak Approximation of Stochastic Equations, *to appear*

[6] T. FUNAKI: Random Motions of Strings and Related Stochastic Evolution Equations, *Nagoya Math. J. 89 (1983) 129-193*

[7] K. IWATA: An Infinite Dimensional Stochastic Differential Equation with State Space $\mathbb{C}(\mathbb{R})$, *Prob. Th. Rel.Fields 74 (1987) 141-159*

[8] R. MANTHEY: Weak Convergence of Solutions of the Heat Equation with Gaussian Noise, *Math. Nachr. 123 (1985) 157-168*

[9] R. MANTHEY: On the Cauchy Problem for Reaction-Diffusion Equations with White Noise, *Math. Nachr. 136 (1988) 209-228*

[10] R. MANTHEY: Reaktions-Diffusionsgleichungen mit weißem Rauschen, *Diss.B, Friedrich-Schiller-Universitat, Jena, 1988*

[11] R. MANTHEY: On the Notion of Solution to Stochastic Partial Differential Equations with Additive White Noise, *submitted*

[12] R. MANTHEY, B. MASLOWSKI: Qualitative Behaviour of Solutions to Stochastic Reaction-Diffusion Equations, *to appear*

[13] J.B. WALSH: A Stochastic Model of Neural Response, *Adv.Appl. Prob. 13 (1981) 231-281*

[14] J.B. WALSH: An Introduction to Stochastic Partial Differential Equations, *Lecture Notes in Mathematics, Vol. 1180, Springer, 1987*

International Series of Numerical Mathematics, Vol. 102, © 1991 Birkhäuser Verlag Basel

SMALL PERTURBATIONS FOR QUASILINEAR ANTICIPATING STOCHASTIC DIFFERENTIAL EQUATIONS *

A. Millet[1] *D. Nualart*[2] and *M. Sanz*[2]

1) Faculté des Sciences. Université d'Angers
2, Bd Lavoisier. 49045 ANGERS. France.
and
Laboratoire de Probabilités
Université de Paris VI (UA 224)
2) Facultat de Matemàtiques. Universitat de Barcelona.
Gran Via 585. 08007 BARCELONA. Spain.

In the last years there have been several attempts to study stochastic integrals with respect to the Brownian motion involving non necessarily adapted stochastic processes. In particular an anticipating stochastic calculus concerning the Skorohod integral has been developed by Nualart and Pardoux in [3]. On the basis of this calculus one can study stochastic differential equations whose solutions are non adapted processes (see [4] for a survey).

In this paper we deal with a special kind of these equations introduced by Buckdahn in [2]. More precisely, we consider the one dimensional stochastic differential equation in the Skorohod sense, defined on the canonical Wiener space,

$$X_t = X_0 + \int_0^t \sigma X_s dW_s + \int_0^t b(X_s)\, ds, \tag{0.1}$$

where $\{W_s, s \in [0,1]\}$ is a Brownian motion, $t \in [0,1]$ and X_0 is a random variable depending on the whole path of W.

Under suitable conditions on X_0 and b (see Theorem 1.2) Buckdahn has proved the existence and uniqueness of solution using Girsanov's transformation. Moreover, if X_0 is smooth enough the solution X_t has almost surely continuous paths (see Proposition 1.3).

Let us now consider the perturbed family

$$X_t^\varepsilon = X_0^\varepsilon + \sqrt{\varepsilon} \int_0^t \sigma X_s^\varepsilon dW_s + \int_0^t b(X_s^\varepsilon)\, ds\,,$$

$\varepsilon > 0$, and assume that the following condition is satisfied:

(C) There exists $x_0 \in \mathbb{R}$ such that, for any $\eta > 0$

$$\lim_{\varepsilon \to 0} \varepsilon \log P\{|X_0^\varepsilon - x_0| > \eta\} = -\infty\,.$$

* This work has been partially supported by the Grant of the C.I.C.Y.T. n.° PB86-0238.

The purpose of this paper is to show that, under the hypothesis (C), the family $\{P^{\varepsilon}, \varepsilon > 0\}$ of laws of $\{X^{\varepsilon}, \varepsilon > 0\}$ satisfies a large deviation principle on $C([0,1])$ with some rate function I(.). (Theorem 2.1).

1. Quasilinear anticipating stochastic differential equations.

We denote by $(\Omega, \mathcal{F}, P)$ the canonical space associated with the standard Wiener process $\{W_t, t \in [0,1]\}$. That means, Ω is the space of continuous functions on [0,1] vanishing at the origin, P is the standard Wiener measure on Ω and $\mathcal{F}$ is the Borel σ-field on Ω completed with respect to the measure P.
Consider the space $\mathcal{S}$ of random variables of the form

$$F(\omega) = f(\omega_{t_1}, \dots, \omega_{t_n}), \tag{1.1}$$

where $t_1, \dots, t_n \in [0,1]$, $n \geq 1$, and f belongs to $C_b^{\infty}(\mathbb{R}^n)$, the space of infinitely differentiable functions which are bounded together with all their derivatives. The elements of $\mathcal{S}$ are called *smooth functionals.*
The derivative of a smooth functional F of the form (1.1) is the stochastic process $\{D_tF,\ 0 \leq t \leq 1\}$ given by

$$D_tF(\omega) = \sum_{i=1}^{n} \frac{\partial f}{\partial x_i}(\omega_{t_1}, \dots, \omega_{t_n}) \mathbf{1}_{[0,t_i]}(t).$$

For any real number $p \geq 1$ we denote by $\mathbb{D}^{1,p}$ the completion of $\mathcal{S}$ with respect to the norm

$$\|F\|_{1,p} = \|F\|_{L^p(\Omega)} + \| \, \|DF\|_{HS} \|_{L^p(\Omega)},$$

where $\|\cdot\|_{HS}$ is the Hilbert–Schmidt norm, that means, the $L^2([0,1])$ norm.
The operator D is a closed unbounded operator from $L^2(\Omega)$ into $L^2([0,1] \times \Omega)$ whose domain is $\mathbb{D}^{1,2}$. Its adjoint is the *Skorohod integral,* which we will denote by δ.
Hence, the domain of δ is the set of processes $u \in L^2([0,1] \times \Omega)$ such that for any $F \in \mathcal{S}$.

$$\left| E \int_0^1 D_tF u_t dt \right| \leq C \|F\|_{L^2(\Omega)},$$

where C is some constant depending on u, and the Skorohod integral of $u, \delta(u)$, is the square integrable random variable determined by the duality relation

$$E(\delta(u)F) = E\Big(\int_0^1 D_tF u_t dt \Big),$$

for all $F \in \mathcal{S}$.
Let us introduce some classes of processes related with the domain of δ. For any real number $p \geq 1$ we set $\mathbb{L}^{1,p} = L^p([0,1]; \mathbb{D}^{1,p})$. The processes u of $\mathbb{L}^{1,2}$ verify $u\mathbf{1}_{[0,t]} \in \text{Dom}\delta$ for each $t \in [0,1]$. Therefore, for these processes we can define the indefinite Skorohod integral $X_t = \delta(u\mathbf{1}_{[0,t]}) = \int_0^t u_s dW_s$. Moreover, if for some

$p > 2, E\Big(\int_0^1 \big(\int_0^1 |D_s u_t|^2 ds\big)^p dt\Big)$ is finite, then $\{X_t, t \in [0,1]\}$ has a continuous version (see Theorem 5.2 in [3]).
The following lemma will be used in the proof of our main result.

Lemma 1.1. Let $F : \Omega \to \mathbb{R}$ be a functional in $\mathbb{D}^{1,p}$, with $p > 1$ and $h : [0,1] \to \mathbb{R}$ in $L^2([0,1])$. Then,

$$F\Big(\omega + \int_0^{\cdot} h_r \mathbf{1}_{[0,t]}(r) dr\Big) - F(\omega) = \int_0^t D_s F\Big(\omega + \int_0^{\cdot \wedge s} h_r dr\Big) h_s \, ds\,.$$

Proof: Assume first that F is a smooth functional of the from (1.1). Then,

$$\begin{aligned} & f\Big(\omega_{t_1} + \int_0^{t\wedge t_1} h_r dr, \ldots, \omega_{t_n} + \int_0^{t\wedge t_n} h_r dr\Big) - f(\omega_{t_1}, \ldots, \omega_{t_n}) \\ &= \int_0^t \sum_{i=1}^n \frac{\partial f}{\partial x_i}\Big(\omega_{t_1} + \int_0^{s\wedge t_1} h_r dr, \ldots, \omega_{t_n} + \int_0^{s\wedge t_n} h_r dr\Big) h_s \mathbf{1}_{[0,t_i]}(s)\, ds \\ &= \int_0^t D_s F\Big(\omega + \int_0^{s\wedge \cdot} h_r dr\Big) h_s \, ds. \end{aligned}$$

Then the result follows by a standard approximation argument. ∎

Consider the one–dimensional stochastic differential equation defined on the probability space $(\Omega, \mathcal{F}, P)$,

$$X_t = X_0 + \int_0^t \sigma X_s dW_s + \int_0^t b(X_s)\, ds, \tag{1.2}$$

where $t \in [0,1]$, $\sigma \in \mathbb{R}$ and the stochastic integral appearing in (1.2) is a Skorohod integral.
We assume that the following conditions are satisfied:

(H_1) X_0 is a $\mathcal{F}$-measurable random variable which belongs to $L^p(\Omega)$ for some $p > 2$.
(H_2) b is a real Lipschitz function.

Equation (1.2) is a particular case of the quasilinear anticipating stochastic differential equations studied by Buckdahn in [2], for which we can give an explicit solution, as follows.
Consider the family of transformations on the Wiener space $T_t, A_t : \Omega \to \Omega$, $t \in [0,1]$ defined by

$$\begin{aligned} T_t(\omega)_s &= \omega_s + \sigma(t \wedge s), \\ A_t(\omega)_s &= \omega_s - \sigma(t \wedge s). \end{aligned}$$

Notice that $T_t A_t = A_t T_t = Id$.
Set $\psi_t(\omega) = \exp\big\{\sigma\omega_t - \frac{1}{2}\sigma^2 t\big\}$, and denote by $Z_t(\omega, x)$ the solution of

$$Z_t(\omega, x) = x + \int_0^t \psi_s^{-1}(T_t(\omega)) b\big(\psi_s(T_t(\omega)) Z_s(\omega, x)\big)\, ds. \tag{1.3}$$

For $s \le t$ we have $\psi_s(T_t) = \exp\{\sigma\omega_s + \frac{1}{2}\sigma^2 s\}$ and consequently (1.3) is equivalent to

$$Z_t(\omega, x) = x + \int_0^t \exp\{-\sigma\omega_s - \frac{1}{2}\sigma^2 s\}\, b\Big(\exp\{\sigma\omega_s + \frac{1}{2}\sigma^2 s\} Z_s(\omega, x)\Big)\, ds.$$

Theorem 1.2. (Theorem 5.2 [2]). Assume that hypothesis (H_1) and (H_2) are satisfied and define

$$X_t(\omega) = \psi_t(\omega) Z_t\big(A_t(\omega),\, X_0(A_t(\omega))\big). \tag{1.4}$$

Then the process $\{X_t, t \in [0,1]\}$ verifies $\mathbf{1}_{[0,t]}X \in \mathrm{Dom}\delta$ for any $t \in [0,1]$, $X \in L^2([0,1]\times\Omega)$ and X is the unique solution of equation (1.2) satisfying these conditions.

We want to show that under some additional conditions the process $\{X_t, t \in [0,1]\}$ defined by (1.4) possesses a continuous version.

Proposition 1.3. Assume that X_0 belongs to $\mathbb{D}^{1,p}$ for some $p > 1$. Then the solution of (1.2) has an almost surely continuous modification.

Proof: Clearly, it suffices to check that the process $\Big\{Z_t\big(A_t(\omega),\, X_0(A_t(\omega))\big),\ t \in [0,1]\Big\}$ has an almost surely continuous version.
Notice that

$$Z_t(\omega, x) = x + \int_0^t \psi_s^{-1}\big(T_s(\omega)\big)\, b\big(\psi_s(T_s(\omega)) Z_s(\omega, x)\big)\, ds\,.$$

Therefore $\{Z_t(\omega, x),\ t \in [0,1]\}$ is an adapted process, and consequently

$$Z_t(A_t(\omega), x) = x + \int_0^t \psi_s^{-1}(\omega)\, b\big(\psi_s(\omega) Z_s\big(A_s(\omega), x\big)\big)\, ds. \tag{1.5}$$

From (1.5) it follows easily that the process $\{Z_t\big(A_t(\omega), x\big),\ t \in [0,1],\ x \in \mathbb{R}\}$ possesses a version which is jointly continuous in (t, x).
On the other hand, an application of Lemma 1.1 to $h \equiv -\sigma$ yields that, if $t \le t'$

$$X_0\big(A_{t'}(\omega)\big) - X_0\big(A_t(\omega)\big) = \int_{t'}^t D_s X_0\big(A_s(\omega)\big)\, \sigma\, ds\,.$$

This concludes the proof of the proposition.

2. Large deviations for quasilinear anticipating stochastic differential equations.

In this section we consider the perturbed family of anticipating stochastic differential equations

$$X_t^\varepsilon = X_0^\varepsilon + \sqrt{\varepsilon}\int_0^t \sigma X_s^\varepsilon dW_s + \int_0^t b(X_s^\varepsilon)\, ds\,, \tag{2.1}$$

$\varepsilon > 0$.
We assume that, for each $\varepsilon > 0$, X_0^ε belongs to $\mathbb{D}^{1,p}$ for some $p > 1$. We also suppose that b is a $\mathcal{C}^1$ function with bounded derivative and that the stochastic integral appearing in (2.1) is a Skorohod integral. Therefore by Theorem 1.2 we have

$$X_t^\varepsilon = \psi_t^\varepsilon Z_t^\varepsilon(A_t^\varepsilon, X_0^\varepsilon(A_t^\varepsilon)), \tag{2.2}$$

where

$$\psi_t^\varepsilon(\omega) = \exp\left\{\sqrt{\varepsilon}\sigma\omega_t - \frac{1}{2}\varepsilon\sigma^2 t\right\},$$
$$A_t^\varepsilon(\omega)_s = \omega_s - \sqrt{\varepsilon}\sigma(t \wedge s),$$

and

$$Z_t^\varepsilon(\omega, x) = x + \int_0^t \exp\left\{-\sqrt{\varepsilon}\sigma\omega_s - \frac{1}{2}\varepsilon\sigma^2 s\right\} b\Big(\exp\left\{\sqrt{\varepsilon}\sigma\omega_s + \frac{1}{2}\varepsilon\sigma^2 s\right\} Z_s^\varepsilon(\omega, x)\Big)\, ds.$$

Furthermore, the process $\{X_t^\varepsilon, t \in [0,1]\}$ has, for every $\varepsilon > 0$, an almost surely continuous version (see Proposition 1.3).
Our aim is to prove that, under suitable hypothesis, the family $\{P^\varepsilon, \varepsilon > 0\}$ of probabilities on $C([0,1])$ given by the laws of $\{X^\varepsilon, \varepsilon > 0\}$ obeys the large deviation principle with some rate function I (.).
We denote by H_{x_0} the set of functions $f : [0,1] \to \mathbb{R}$ such that $f(0) = x_0$, $f_t = \int_0^t \dot{f}_s\, ds$ and $\int_0^1 |\dot{f}_s|^2\, ds < +\infty$.
Here is our main result.

Theorem 2.1. Let $\{X^\varepsilon, \varepsilon > 0\}$ be the family of continuous processes solution of the stochastic differential equations (2.1). Assume that the following conditions hold:

(a) There exists $x_0 \in \mathbb{R}$ such that, for any $\eta > 0$

$$\lim_{\varepsilon \downarrow 0} \varepsilon \log P\{|X_0^\varepsilon - x_0| > \eta\} = -\infty.$$

(b) For any $M > 0$ there exists $\varepsilon_0 > 0$ such that

$$\sup_{0<\varepsilon\leq\varepsilon_0} E\Big\{\int_0^1 \exp\{M|D_s X_0^\varepsilon|^2\}\, ds\Big\} < +\infty.$$

Then, the family $\{P^\varepsilon, \varepsilon > 0\}$ of laws of $\{X^\varepsilon, \varepsilon > 0\}$ satisfies the large deviation principle with rate function

$$I(f) = \begin{cases} +\infty, & \text{if } f \notin H_{x_0}, \\ \dfrac{1}{2}\displaystyle\int_0^1 \frac{1}{\sigma^2}\big(\dot{f}_t - b(f_t)\big)^2 \frac{1}{f_t^2}\, dt, & \text{if } f \in H_{x_0}. \end{cases} \tag{2.3}$$

Before giving the proof of this lemma we will state an auxiliary result.

Lemma 2.2. Let (S, d) be a complete separable metric space and $\{P^\varepsilon, \varepsilon > 0\}$ a family of probability measures on the Borel subsets of S. Assume that $\{P^\varepsilon, \varepsilon > 0\}$ obeys the large deviation principle with a rate function I (.). Fix $f_0 \in S$ and define

$F_M(f_0) = \{f \in S : d(f, f_0) \geq M\}$, $M \geq 0$. Then for any $C > 0$ there exists $\varepsilon_0 > 0$ and $M_0 > 0$ such that for each $\varepsilon \leq \varepsilon_0$ and $M \geq M_0$ we have

$$P^\varepsilon(F_M(f_0)) \leq \exp\left\{-\frac{C}{\varepsilon}\right\}.$$

Proof: The sets $F_M(f_0)$, $M \geq 0$ are closed, therefore

$$\overline{\lim_{\varepsilon\to 0}}\, \varepsilon \log P^\varepsilon(F_M(f_0)) \leq -\inf_{f\in F_M(f_0)} I(f).$$

Consequently, for any $\alpha > 0$ there exists $\varepsilon_0 > 0$ such that for any $\varepsilon \leq \varepsilon_0$

$$P^\varepsilon(F_M(f_0)) \leq \exp\left\{\frac{1}{\varepsilon}\Big(-\inf_{f\in F_M(f_0)} I(f) + \alpha\Big)\right\}.$$

Notice that $F_M(f_0) \downarrow \emptyset$ as M tends to infinity.
Fix $C > 0$, we will show that we can choose M_0 such that $\inf_{f\in F_{M_0}(f_0)} I(f) > C$.
Indeed, if not, then there exists $C > 0$ and for any $M > 0$ there exists $f_M \in F_M(f_0)$ such that $I(f_M) \leq C$.
Let $K = \{f \in S : I(f) \leq C\}$. By the properties of the rate functions we know that K is a compact set. Hence, we can find a subsequence $f_{M_i} \in F_{M_i}(f_0) \cap K$ and $f \in K$ such that

$$d(f_{M_i}, f) \underset{i\to\infty}{\longrightarrow} 0.$$

This leads to a contradiction. Indeed, $d(f_{M_i}, f_0) \to \infty$ as i tends to ∞, and on the other hand

$$d(f_{M_i}, f_0) \leq d(f_{M_i}, f) + d(f, f_0).$$

This finishes the proof of the lemma. ■

Proof of theorem 2.1. Let us consider the analogue of (2.1) with a deterministic initial condition $x_0 \in \mathbb{R}$,

$$\xi_t^\varepsilon = x_0 + \sqrt{\varepsilon}\int_0^t \sigma\xi_s^\varepsilon dW_s + \int_0^t b(\xi_s^\varepsilon)\, ds. \tag{2.4}$$

Equation (2.4) is a particular case of (2.1). Consequently, using (2.2) we obtain

$$\xi_t^\varepsilon = \psi_t^\varepsilon Z_t^\varepsilon(A_t^\varepsilon, x_0).$$

Notice that by the classical results about small perturbations of dynamical systems we know that $I(f)$ defined by (2.3) is the rate function corresponding to $\{\xi^\varepsilon, \varepsilon > 0\}$ (see for instance [1], [5]).
The proof of the theorem will be done in two steps.

Step 1. We will first prove that, for any $\eta > 0$,

$$\lim_{\varepsilon\to 0} \varepsilon \log P\{\sup_{0\leq t\leq 1} |X_t^\varepsilon - \xi_t^\varepsilon| > \eta\} = -\infty.$$

That means, we have to check for any $\eta > 0$ and $C > 0$ there exists $\varepsilon_0 > 0$ such that for any $\varepsilon \le \varepsilon_0$

$$P\{\sup_{0\le t\le 1} |X_t^\varepsilon - \xi_t^\varepsilon| > \eta\} \le e^{-\frac{C}{\varepsilon}}. \qquad (2.5)$$

For any $\eta > 0$ and $K > 0$ we have

$$P\{\sup_{0\le t\le 1} |X_t^\varepsilon - \xi_t^\varepsilon| > \eta\} \le P\{\sup_{0\le t\le 1} |\psi_t^\varepsilon| > K\} + P\{\sup_{0\le t\le 1} |\sqrt{\varepsilon}W_t| > K\}$$
$$+ P\{\sup_{0\le t\le 1} |X_t^\varepsilon - \xi_t^\varepsilon| > \eta,\ \sup_{0\le t\le 1} |\psi_t^\varepsilon| \le K,\ \sup_{0\le t\le 1} |\sqrt{\varepsilon}W_t| \le K\}.$$

Notice that the process $\{\psi_t^\varepsilon,\, t \in [0,1]\}$ is the solution of the stochastic differential equation

$$\psi_t^\varepsilon = 1 + \sqrt{\varepsilon}\int_0^t \sigma\psi_s^\varepsilon dW_s,$$

and therefore $\{\psi^\varepsilon, \varepsilon > 0\}$, as well as $\{\sqrt{\varepsilon}W, \varepsilon > 0\}$, satisfies the large deviation principle.
Consequently, using Lemma 2.2, we conclude that for any $C > 0$ there exists $\varepsilon_0 > 0$ and $K_0 > 0$ such that for any $\varepsilon \le \varepsilon_0$ and $K \ge K_0$

$$P\{\sup_{0\le t\le 1} |\psi_t^\varepsilon| > K\} + P\{\sup_{0\le t\le 1} |\sqrt{\varepsilon}W_t| > K\} \le e^{-\frac{C}{\varepsilon}}.$$

By Gronwall's lemma,

$$|Z_t^\varepsilon(\omega, x) - Z_t^\varepsilon(\omega, x')| \le e^{\|b'\|_\infty}|x - x'|,$$

for any $x, x' \in \mathbb{R}$.
Then, if we denote by B the set $\{\sup_{0\le t\le 1} |\psi_t^\varepsilon| \le K,\ \sup_{0\le t\le 1} |\sqrt{\varepsilon}W_t| \le K\}$,
we have

$$P\Big(\{\sup_{0\le t\le 1} |X_t^\varepsilon - \xi_t^\varepsilon| > \eta\} \cap B\Big)$$
$$= P\Big(\{\sup_{0\le t\le 1} |\psi_t^\varepsilon|\,|Z_t^\varepsilon(A_t^\varepsilon, X_0^\varepsilon(A_t^\varepsilon)) - Z_t^\varepsilon(A_t^\varepsilon, x_0)| > \eta\} \cap B\Big)$$
$$\le \Big(P\{\sup_{0\le t\le 1} |X_0^\varepsilon(A_t^\varepsilon) - x_0| > \eta'\} \cap B\Big),$$

with $\eta' = \dfrac{\eta}{K}e^{-\|b'\|_\infty}$.
Due to hypothesis (a), in order to check (2.5) it suffices to estimate

$$\alpha = P\Big(\{\sup_{0\le t\le 1} |X_0^\varepsilon(A_t^\varepsilon) - X_0^\varepsilon| > \frac{\eta'}{2}\} \cap B\Big).$$

We now apply Lemma 1.1 to $F = X_0^\varepsilon$ and $h_r = -\sigma\sqrt{\varepsilon}$ and we obtain

$$\alpha = P\Big(\{\sup_{0\le t\le 1} \Big|\int_0^t D_sX_0^\varepsilon(A_s^\varepsilon(\omega))\sigma\sqrt{\varepsilon}\,ds\Big| > \frac{\eta'}{2}\} \cap B\Big).$$

Therefore, for any $M > 0$

$$\alpha \le P\Big(\Big\{\exp M\Big(\int_0^1 \big|D_s X_0^\varepsilon(A_s^\varepsilon(\omega))\big|\, ds\Big)^2 > e^{\frac{M\beta}{\varepsilon}}\Big\} \cap B\Big),$$

with $\beta = \left(\frac{\eta'}{2\sigma}\right)^2$, and by Chebychev and Jensen's inequalities

$$\alpha \le e^{-\frac{M\beta}{\varepsilon}} \int_B \exp M\Big(\int_0^1 \big|D_s X_0^\varepsilon(A_s^\varepsilon(\omega))\big| ds\Big)^2 dP$$
$$\le e^{-\frac{M\beta}{\varepsilon}} \int_0^1 \Big(\int_B \exp M\big|D_s X_0^\varepsilon(A_s^\varepsilon(\omega))\big|^2 dP\Big)\, ds.$$

By Girsanov's theorem we know that $A_s^\varepsilon(\omega)$ is a Wiener process under $\psi_s^\varepsilon P$. Hence,

$$\alpha \le e^{-\frac{M\beta}{\varepsilon}} \int_0^1 \Big(\int_{\{\sup_{0\le s\le 1} |\sqrt{\varepsilon}\omega_s| \le k'\}} \exp M\big|D_s X_0^\varepsilon(\omega)\big|^2 \exp\big\{-\sigma\sqrt{\varepsilon}\,\omega_s - \frac{1}{2}\sigma^2 \varepsilon s\big\} dP\Big) ds,$$

for some constant $k' > 0$.
Consequently, hypothesis (b) allows us to conclude that $\alpha \le C_0 e^{-\frac{M\beta}{\varepsilon}}$, for some constant C_0, and the proof of the estimate (2.5) is now complete

Step 2. We want to prove that, for any open set G and any closed set F in $C([0,1])$

$$\underline{\lim}_{\varepsilon\to 0} \varepsilon \log P^\varepsilon(G) \ge -\inf_{f\in G} I(f), \tag{2.6}$$

and

$$\overline{\lim}_{\varepsilon\to 0} \varepsilon \log P^\varepsilon(F) \le -\inf_{f\in F} I(f). \tag{2.7}$$

These estimations will be obtained using the idea of the proof of Theorem 6.3 in [5]. For the sake of completeness we will give some details.
Fix G an open set in $C([0,1])$. Without loss of generality we may assume that $G \cap H_{x_0} \ne \emptyset$. Let $f \in G \cap H_{x_0}$; we fix $\eta > 0$ and a neighbourhood U of f such that

$$U_\eta = \{g \in C([0,1]) : d(g,U) < \eta\} \subset G.$$

Then we can write

$$P^\varepsilon(G) = P\{X^\varepsilon \in G\} \ge P\{\xi^\varepsilon \in U\} - P\{\sup_{0\le t\le 1} |X_t^\varepsilon - \xi_t^\varepsilon| > \eta\}.$$

Since $I(\cdot)$ is the rate function for the laws of $\{\xi^\varepsilon, \varepsilon > 0\}$, we obtain

$$\underline{\lim}_{\varepsilon\to 0} \varepsilon \log P\{\xi^\varepsilon \in U\} \ge -\inf_{f\in U} I(f). \tag{2.8}$$

Then (2.8) and (2.5) provide the lower bound (2.6).
Let us now check the upper bound (2.7). Fix $\eta > 0$ and consider the sets $F_\eta = \{g \in C([0,1]) : d(g,F) < \eta\}$. We denote by $\overline{F_\eta}$ the closure of F_η.

Then,

$$P^\varepsilon(F) = P\{X^\varepsilon \in F\} \le P\{\xi^\varepsilon \in \overline{F_\eta}\} + P\{\sup_{0\le t\le 1} |\xi_t^\varepsilon - X_t^\varepsilon| \ge \eta\}\,.$$

We have

$$\overline{\lim_{\varepsilon\to 0}}\, \varepsilon \log P\{\xi^\varepsilon \in \overline{F_\eta}\} \le -\inf_{f\in\overline{F_\eta}} I(f)\,.$$

This property, together with the estimation (2.5), imply that for any $C > 0$ and $\alpha > 0$ there exists $\varepsilon_1 > 0$ such that if $0 < \varepsilon \le \varepsilon_1$,

$$P^\varepsilon(F) \le \exp \frac{1}{\varepsilon}\big[\alpha - \inf_{f\in\overline{F_\eta}} I(f)\big] + \exp\Big(-\frac{C}{\varepsilon}\Big)\,.$$

Therefore

$$\overline{\lim_{\varepsilon\to 0}}\, \varepsilon \log P^\varepsilon(F) \le -\inf_{f\in\overline{F_\eta}} I(f)\,. \tag{2.9}$$

Since the left hand side of (2.9) does not depend on η, and $\overline{F_\eta} \downarrow F$ as $\eta \downarrow 0$, we obtain (2.7), and this ends the proof of the theorem. ∎

References

1. Azencott, R.: Grandes déviations et applications. Ecole d'été de Probabilités de Saint-Flour VIII-1978. Lecture Notes in Math. 774. Springer Verlarg. Berlin-Heidelberg-NewYork 1980.
2. Buckdahn, R.: Quasilinear partial differential equations without nonanticipation requirement. Sektion Mathematik der Humboldt-Universität. Preprint 176. Berlin, 1988.
3. Nualart, D. and Pardoux, E.: Stochastic calculus with anticipating integrands. Probab. Th. Rel. Fields 78, 535–581 (1988).
4. Pardoux. E.: Applications of anticipating Stochastic Calculus to stochastic differential equations. To appear in H. Korezlioglu, A. S. Ustunel (Eds.): Stochastic Analysis and Related Topics. Proceedings, Silivri 1986. Lecture Notes in Math. Springer Verlag.
5. Varadahn, S. R. S.: Large deviations and Applications. CBMS–NSF regional conference series in applied Mathematics, 46. SIAM. Philadelphia 1984.

This work was carried out while the first author was visiting the "Centre de Recerca Matemàtica" at Barcelona.

SIMULATION AND ANALYSIS OF MECHANICAL SYSTEMS WITH PARAMETER FLUCTUATIONS

Walter V. Wedig
Institut für Technische Mechanik, Universität Karlsruhe
Kaiserstr. 12, D - 7500 Karlsruhe 1

Abstract

Mechanical systems with parameter fluctuations lead to bifurcation problems. With increasing fluctuation intensity the equilibrium of the system becomes unstable and bifurcates into non-trivial stationary solutions. To illustrate this effect we consider the typical example of a simply supported beam under axial loading. Its governing partial differential equation of motion can be reduced to a non-linear ordinary one. For the special case of white noise fluctuations we investigate Lyapunov exponents and bifurcation points of both solution forms.

1. Bifurcation problems of mechanical systems

As a representative example of bifurcations in mechanical systems, we consider a uniform beam under axial loading $P(t)$. According to figure 1, the beam has the length l, the bending stiffness EI, the axial siffness EA, the mass μ per unit length and the external viscous damping β. Following Weidenhammer (1969), the transverse motion $w(x,t)$ of a beam section is described by a boundary value problem, given as follows:

$$EIw_{xxxx} + \beta w_t + \mu w_{tt} - \tfrac{1}{l}EA[u(t) + \tfrac{1}{2}\int_0^l w_{xx}^2 dx]w_{xx} = 0, \tag{1}$$

$$w(0,t) = w(l,t) = w_{xx}(0,t) = w_{xx}(l,t) = 0, \qquad 0 \le x \le l. \tag{2}$$

Herein, sub t and sub x denote partial derivations of $w(x,t)$ with respect to the time t and to the length coordinate x, respectively. The governing equation of motion (1) is a non-linear integro-differential equation which provides Kirchhoff's assumption that longitudinal waves $u(x,t)$ are negligible. Correspondingly, $u(t)$ represents only the end deflection of the beam determined by the axial force $P(t)$.

As illustrated in figure 1, we restrict our interest to the special case of simple hinged supports that the displacements $w(x,t)$ and moments $EIw_{xx}(x,t)$ at both ends of the beam possess the homogeneous boundary conditions (2) valid for each time $t \ge 0$. Obviously, these conditions are satisfied by the first mode $w(x,t) = T(t)\sin \pi x/l$ which is associated to the first natural frequency ω_1 of the unloaded and undamped beam.

$$w(x,t) = T(t)\sin \pi x/l, \qquad \omega_1^2 = \pi^4 EI/(\mu l^4), \tag{3}$$

$$\ddot{T}(t) + 2D\omega_1\dot{T}(t) + \omega_1^2[1 + \frac{Al}{I\pi^2}u(t) + \gamma T^2(t)]T(t) = 0. \tag{4}$$

Insertion of the first mode reduces the partial differential equation (1) to an ordinary differential equation which is noted in (4). Herein, the parameter γ characterizes the cubic restoring of the beam and D denotes a dimensionless damping value. Both parameters are calculabe by inspection.

The equation (4) possesses two different solutions: the trivial solution $T(t) \equiv 0$ and the bifurcated one $T^2(t) > 0$. The stability of both solution forms is decided by the top Lyapunov exponents λ_0 and λ_S, respectively. Herein, the index 0 denotes the top Lyapunov exponent of the trivial solution and correspondingly, the index S belongs to the stationary bifurcation. Negative Lyapunov exponents λ_0 determine an exponentially decreasing of perturbed solutions to the equilibrium. In the present paper, we restrict our interest to λ_0.

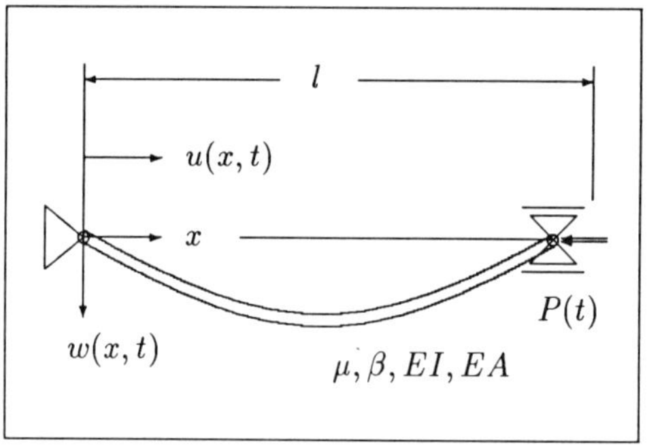

Figure 1: Beam under axial loading

2. Simulation of stochastic bifurcation problems

As already mentioned, we are interested in the special case that the parameter fluctuation $Alu(t)/(\pi^2 I)$ can be modelled by stationary white noise $\sigma\dot{W}_t$ where σ is the intensity and $\dot{W}_t$ is normalized with zero mean and the delta-correlation $E(\dot{W}_t\dot{W}_s) = \delta(t-s)$. Consequently, $T(t)$ is a stochastic process represented by X_t, and the bifurcation equation (4) reads as follows:

$$\ddot{X}_t + 2D\omega_1\dot{X}_t + \omega_1^2\left[1 + \sigma\dot{W}_t + \gamma X_t^2\right]X_t = 0. \tag{5}$$

Following the concept of Khasminskii (1967), we introduce polar coordinates via the amplitude process A_t and the phase process Ψ_t

$$X_t = A_t\cos\Psi_t, \qquad A_t = \sqrt{X_t^2 + (\dot{X}_t/\omega_1)^2}, \tag{6}$$

$$\dot{X}_t = \omega_1 A_t\sin\Psi_t, \qquad \Psi_t = \arctan\dot{X}_t/(\omega_1 X_t) \tag{7}$$

and transform the second order equation (5) into a first order system of stochastic Itô differential equations for the increments of A_t and Ψ_t.

$$\begin{aligned} dA_t = &-2D\omega_1 A_t\sin^2\Psi_t dt + \tfrac{1}{2}\omega_1^2\sigma^2 A_t\cos^4\Psi_t dt \\ &-\omega_1\gamma A_t^3\sin\Psi_t\cos^3\Psi_t dt - \omega_1\sigma A_t\sin\Psi_t\cos\Psi_t dW_t, \end{aligned} \tag{8}$$

$$\begin{aligned} d\Psi_t = &-\omega_1(1 + 2D\sin\Psi_t\cos\Psi_t)dt - \omega_1\sigma\cos^2\Psi_t dW_t \\ &-\omega_1\gamma A_t^2\cos^4\Psi_t dt - \omega_1^2\sigma^2\sin\Psi_t\cos^3\Psi_t dt. \end{aligned} \tag{9}$$

Herein, W_t is the normed Wiener process. It is normally distributed with zero mean and uncorrelated increments. The $(\omega_1\sigma)^2$- terms in both equations result from the application of Itô's calculus. More details of this calculus are given e.g. by Arnold (1974). Finally, we introduce the natural logarithm $L_t = \log A_t$ of the amplitude process and calculate its increment.

$$d\log A_t = -2D\omega_1 \sin^2 \Psi_t dt - \omega_1\gamma A_t^2 \sin \Psi_t \cos^3 \Psi_t dt \\ -\tfrac{1}{2}\omega_1^2\sigma^2 \cos^2 \Psi_t(\sin^2 \Psi_t - \cos^2 \Psi_t)dt - \omega_1\sigma \sin \Psi_t \cos \Psi_t dW_t. \quad (10)$$

This form is needed for the determination of Lyapunov exponents derived in the last section of the paper.

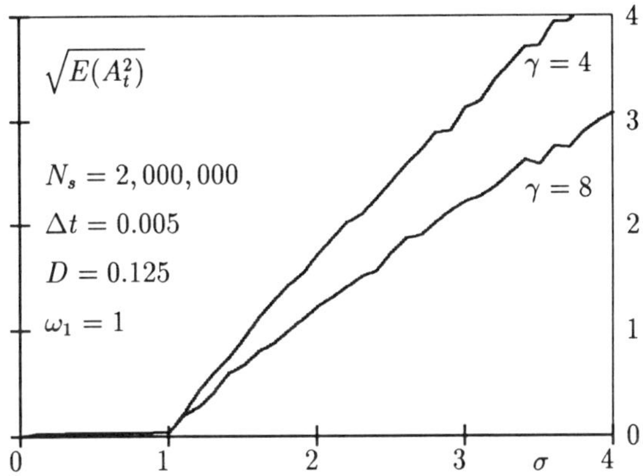

Figure 2: Simulated bifurcation diagram

Before going into more details of the stability analysis we perform simple numerical experiments to show basic effects of the bifurcation problem of interest. The realized simulation routine is based on the forward Euler scheme applied to the phase equation (9) and to the log-amplitude equation (10).

$$\Psi_{n+1} = \Psi_n - \omega_1\Delta t(1 + 2D\sin\Psi_n\cos\Psi_n) - \omega_1\sigma\sqrt{\Delta_t}R_n\cos^2\Psi_n \\ -\omega_1\gamma\Delta t\exp(2L_n)\cos^4\Psi_n - \omega_1^2\sigma^2\Delta t\sin\Psi_n\cos^3\Psi_n, \quad (11)$$

$$L_{n+1} = L_n - 2D\omega_1\Delta t\sin^2\Psi_n - \tfrac{1}{2}\omega_1^2\sigma^2\Delta t\cos^2\Psi_n(\sin^2\Psi_n - \cos^2\Psi_n) \\ -\omega_1\sigma\sqrt{\Delta t}R_n\sin\Psi_n\cos\Psi_n - \omega_1\gamma\Delta t\exp(2L_n)\sin\Psi_n\cos^3\Psi_n. \quad (12)$$

Herein, Δt denotes the time step. The Ψ_n and L_n are time-discrete versions of the phase process Ψ_t and of the log-amplitude process L_t. R_n is a sequence of uncorrelated numbers which are normally distributed with zero mean and $E(R_n^2) = 1$. This corresponds to the time-continuous versions $E(dW_t) = 0$ and $E(dW_t^2) = dt$. Figure 2 shows numerical results obtained for the system data $\omega_1 = 1$, $D = 0.125$ and for the two cubic restoring parameters $\gamma = 4$ and $\gamma = 8$. The time step selected was $\Delta t = 0.005$ applied for $N_s = 2,000,000$ sample points. The recursion formulas (11) and (12) produced a stationary sequence of L_n-values from which we calculated the time-discrete amplitudes A_n. Squaring them and summing up we computed the time average and plotted the associated root mean square against the intensity parameter σ. The plots show clearly that there is a critical intensity parameter or bifurcation point σ_c where the trivial solution $A_t \equiv 0$ bifurcates into a stationary solution $A_t^2 > 0$. In particular, we recognize that the bifurcation point is not influenced by the non-linear parameter γ.

3. Invariant measure of the stationary phase

To investigate the stability of the trivial solution $A_t \equiv 0$, the amplitude equation (8) and the phase equation (9) are linearized with respect to $A_t << 1$. This leads to an uncoupled phase equation of the following form:

$$d\Psi_t = -\omega_1[1 + (2D + \omega_1\sigma^2 \cos^2 \Psi_t) \sin \Psi_t \cos \Psi_t]dt - \omega_1\sigma \cos^2 \Psi_t dW_t. \tag{13}$$

By means of the forward difference scheme applied in (11), the phase process is simulated for $\omega_1 = 1, D = 0.125$ and for the two different excitation intensities $\sigma = 1$ and $\sigma = 3$. In figure 3, we show histograms of the estimated phase densities $p(\psi)$ obtained for a step size $\Delta t = 0.005$ and $N_s = 1,000,000$ sample points. From this it follows that the phase process is stationary. It rotates clockwise mainly driven by the inhomogeneous drift term $-\omega_1 dt$. For the special angles $\psi = \pi/2 \pm k\pi$ $(k = 0, 1, 2, ..)$, the diffusion term in (13) is vanishing so that the estimations of $p(\psi)$ converge rapidly in these singular points. Since the equation (13) is doubly periodic with respect to ψ, the phase density $p(\psi)$ needs only to be evaluated in the limited angle range $-\pi/2 \leq \psi \leq +\pi/2$.

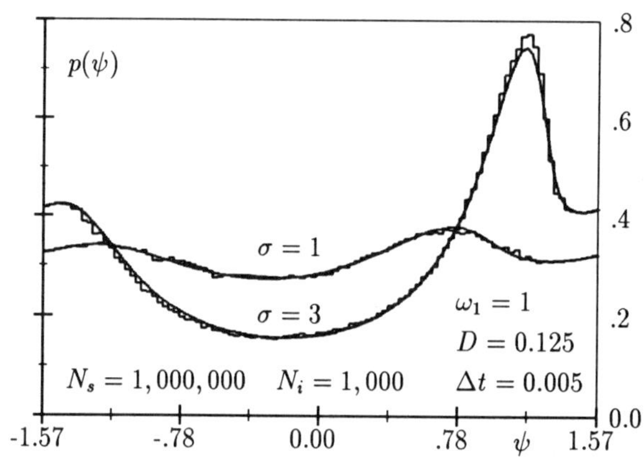

Figure 3: Invariant measure - stationary density

The Itô equation (13) defines a Markov process which is equivalently described by the associated Fokker-Planck equation. In the stationary case, this diffusion equation can be integrated with respect to the angle variable ψ leading to a linear first order differential equation for the phase density $p(\psi)$.

$$\tfrac{1}{2}\omega_1\sigma^2 \cos^4 \psi \, p'(\psi) + [1 + (2D - \omega_1\sigma^2 \cos^2 \psi) \sin \psi \cos \psi]p(\psi) = C, \tag{14}$$

$$-\pi/2 \leq \psi \leq +\pi/2, \qquad p(\pm\pi/2) = C, \qquad \int_{-\pi/2}^{+\pi/2} p(\psi)d\psi = 1. \tag{15}$$

Herein, C is a constant of integration to be determined by the normalization condition, noted in (15). Derivatives with respect to ψ are denoted by primes. Obviously, the Fokker-Planck equation (14) possesses two singularities at the boundaries $\psi = \pm\pi/2$. To select a regular physically meaningful solution $p(\psi)$, we can apply a singular perturbation method (see Wedig, 1988) for expansions around the two singular points. Inside the regular angle range, we can map the linear solution obtained by classical integration routines.

According to Wedig (1989), the method, mentioned above, can be simplified considerably by means of a backward difference scheme which regularize the numerical integration in (14). It leads to the following recurrence formula:

$$p'(\psi) = (p_n - p_{n-1})/\Delta\psi, \quad (n = 1, 2, ...N), \qquad \Delta\psi = \pi/N, \tag{16}$$

$$p_n = \frac{2\Delta\psi C + \omega_1\sigma^2\cos^4\psi_n p_{n-1}}{2\Delta\psi\,[1 + (2D - \omega_1\sigma^2\cos^2\psi_n)\sin\psi_n\cos\psi_n] + \omega_1\sigma^2\cos^4\psi_n}. \tag{17}$$

The recursive computation in (17) begins with $\psi_0 = -\pi/2$ and $p_0 = p(-\pi/2) = C$. It ends at the upper angle with the same density value $p_N = p(+\pi/2) = C$. The recursion is numerically stable for sufficiently small step sizes $\Delta\psi = \pi/N$. Figure 3 shows a comparison of the density histograms simulated by (13) for $N_s = 1{,}000{,}000$ sample points with the smooth lines, calculated from (17) for the interval number $N_i = 1{,}000$.

4. Lyapunov exponents of the trivial solution

The stationary phase density $p(\psi)$ represents an invariant measure of the linearized system equation (5). It determines the exponentially growth behaviour of the associated state vector $\vec{X}_t$. This growth can be investigated by determining the top Lyapunov exponent λ_0 which is to be calculated from the linearized amplitude equation (8).

$$\begin{aligned} d\log A_t = &-\omega_1\sigma\sin\Psi_t\cos\Psi_t dW_t - \omega_1[2D\sin^2\Psi_t \\ &+\tfrac{1}{2}\omega_1\sigma^2\cos^2\Psi_t(\sin^2\Psi_t - \cos^2\Psi_t)]dt. \end{aligned} \tag{18}$$

Invoking the multiplicative ergodic theorem of Oseledec (1968) the top Lyapunov exponent λ_0 is obtained as

$$\lambda_0 = \lim_{t\to\infty}\tfrac{1}{t}\log \| \vec{X}_t \| \,/\, \| \vec{X}_0 \| = \lim_{t\to\infty}\tfrac{1}{t}\log(A_t/A_0), \tag{19}$$

where $\| \cdot \|$ denotes a suitable norm of the system vector $\vec{X}_t$. It is suitable to take the Euclidean norm, already introduced by the amplitude process A_t. The A_0 and $\| \vec{X}_0 \|$ represent associated initial conditions at the time $t_0 = 0$.

The linear amplitude equation (18) can easily be integrated leading to the solution

$$\begin{aligned} \log(A_t/A_0) = &-\omega_1\sigma\int_0^t \sin\Psi_\tau\cos\Psi_\tau dW_\tau - \omega_1\int_0^t [2D\sin^2\Psi_\tau \\ &+\tfrac{1}{2}\omega_1\sigma^2\cos^2\Psi_\tau(\sin^2\Psi_\tau - \cos^2\Psi_\tau)]d\tau. \end{aligned} \tag{20}$$

Inserted into the multiplicative ergodic theorem (19), the integrated form (20) produces time averages like

$$\lim_{t\to\infty}\tfrac{1}{t}\int_0^t(\cdot)d\tau = E(\cdot) = \int_{-\pi/2}^{+\pi/2}(\cdot)p(\psi)d\psi, \qquad \lim_{t\to\infty}\tfrac{1}{t}\int_0^t(\cdot)dW_\tau = 0. \tag{21}$$

The time average of the diffusion term is vanishing because of the non-anticipating property of Itô integrals. The time average of the drift term can be replaced by corresponding expected values defined by the invariant measure of the stationary phase process. Consequently, the top Lyapunov exponent λ_0 is given by

$$\lambda_0 = -2D\omega_1 E(\sin^2\Psi_t) - \tfrac{1}{2}\omega_1^2\sigma^2 E[\cos^2\Psi_t(\sin^2\Psi_t - \cos^2\Psi_t)]. \tag{22}$$

It is completely determined by the stationary density distribution (17) of the phase process.

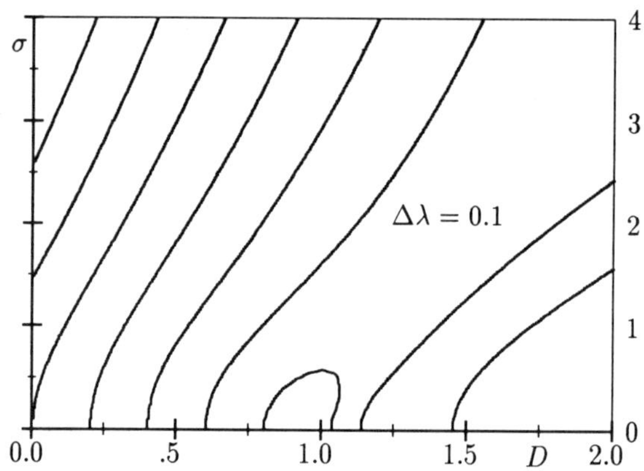

Figure 4: Contour lines of Lyapunov exponents

Figure 4 shows an evaluation of the top Lyapunov exponent λ_0 in the parameter range $0 \leq D \leq 2$ and $0 \leq \sigma \leq 4$ for the natural frequency $\omega_1 = 1$. The calculations are performed by means of the recurrence formula (17) applying Simpson's rule for the integrals of the expectation values in (22). The obtained Lyapunov exponents are given in form of contour lines with the level interval $\Delta\lambda = 0.1$. The line starting with the parameter values $\sigma = D = 0$ denotes the zero level or stability boundary of the linear system. The Lyapunov exponents are negative below this line, and they increase with increasing intensities of the parametric excitation. Right from the stability boundary, the Lyapunov exponents decrease with increasing damping ratios D up to a certain optimal line where they increase again. For the special deterministic case $\sigma = 0$, the Lyapunov exponents, calculated above, coincide with the eigenvalues of the unperturbed system. Further controls are possible in the stability boundary, already derived by Kozin and Mitchel (1974) via singularity methods. The contour lines of figure 4 can also be compared with simulated results of Kliemann and Arnold (1981) for parametric excitations by coloured noise.

References

1. Weidenhammer, F. 1969. Biegeschwingungen des Stabes unter axial pulsierender Zufallslast. VDI-Berichte Nr. 135: 101-107.

2. Khasminskii, R.Z. 1967. Necessary and sufficient conditions for asymptotic stability of linear stochastic systems. Theor. Prob. and Appls. 12: 144-147.

3. Arnold, L. 1974. Stochastic Differential Equations. New York: Wiley.

4. Wedig, W. 1988. Pitchfork and Hopf bifurcations in stochastic systems - effective methods to calculate Lyapunov exponents. To appear in: Effective Stochastic Analysis (ed. by P. Krée, W. Wedig), Heidelberg: Springer.

5. Wedig, W. 1989. Vom Chaos zur Ordnung. Gamm-Mitteilungen ISSN 0936-7195, Heft 2: 3-31.

6. Oseledec, V.I. 1968. A multiplicative ergodic theorem, Lyapunov characteristic numbers for dynamical systems. Trans. Moscow Math. Soc. 19: 197-231.

7. Kozin, F. & Mitchel, R.R. 1974. Sample stability of second order linear differential equations with wide band noise coefficients. SIAM J. Appl. Math., 17: 571-605.

8. Arnold, L. & Kliemann, W. 1981. Qualitative theory of stochastic systems. In: Probabilistic Analysis and Related Topics (ed. by A.T. Bharucha-Reid). Vol. 3, New York: Academic Press.